AF885511

SOTHIS PRESS

Sothis Press aims to completely reimagine the public domain book with an emphasis on universal access to captivatingly designed print editions. We believe that technology will enable every single book ever published to be perpetually accessible in beautiful print form. Sothis is beauty in ubiquity.

www.sothis.press
1 Waterhouse Square, London, ECIN 2ST

Handbook of Explosives for Farmers, Planters, Ranchers

How to Clear Land of Stumps, Trees or Boulders, Dig Ditches, Grade Roads, Excavate Cellars and Foundation Trenches, Dig Holes for Poles and Posts, Break Up Hardpan or Other Hard Soil, Plant and Cultivate Fruit Trees, Start Log Jams and Ice Gorges

Canadian Explosives Limited

SOTHIS PRESS

ISBN 978-1-0365-1388-7

Handbook of Explosives for Farmers, Planters, Ranchers is in the public domain.

This edition is published by Sothis Press © 2023. The company is registered in England and Wales. The Company Number is 14900389.

HAND BOOK OF EXPLOSIVES
for
Farmers-Planters-Ranchers

HOW TO
Clear Land of Stumps, Boulders or Trees
Dig Ditches
Grade Roads
Excavate Cellars and Foundation Trenches
Sink Wells
Dig Holes for Poles and Posts
Break up Hardpan or other Hard Soil
Plant and Cultivate Fruit Trees
Start Log Jams and Ice Gorges
with

DU PONT
EXPLOSIVES

Copyrighted

[illegible]

By

E.I. DU PONT [illegible] POWDER CO.

WIL[illegible] DEL.

INDEX

Bar for Punching Holes 56
Bigtree Stumps 35
Blasting Caps 105
Blasting by Electricity 108
Blasting Machines 109
Blasting Powder Kegs 92
Blasting Supplies 105
Boulder Blasting 45
Branch Offices 120
Burning Out Stumps 19

Caps: Blasting 105
Cap Crimpers 107
Cartridges of High Explosives 94
Cases of High Explosives 92
Cedar Stumps: Western 31
Cellar Digging 63
Charging 103
Clearing Land 13
Connecting Wire 110
Cultivating Fruit Trees 79
Cypress Stumps 39

Detonators 95
Ditching 51
Draining Swamps 60

Electrical Blasting 108
Electric Fuzes 110
Explosives: Principle of 93

Felling Trees 41
Fir Stumps: Western 31
Foundations: Excavating for 63
Fuse 105
Fuzes: Electric 110

Hardpan Blasting 69

Ice Blasting 87
Implements Used in Land Clearing 21
Introduction 7

Judson Powder R. R. P. Bag 94

Kegs of Blasting Powder 92

Leading Wire........ 110
Log Jams........ 90
Log Splitting........ 41

Notes........ 104

Offices: Branch........ 120
Ordering........ 112

Packages of Explosives........ 97
Pine Stumps: Southern........ 27
Pine Stumps: Western........ 31
Planting Fruit Trees........ 79
Plowing with Dynamite........ 69
Post Hole Digging........ 65
Precautions........ 113
Priming........ 101, 111
Principle of Explosives........ 93

Redwood Stumps........ 35
Reliable Blasting Machines........ 109
Road Building........ 63

Second-Growth Stumps........ 31
Sinking Wells........ 64
Southern Pine Stumps........ 27
Splitting Logs........ 41
Storage of Explosives........ 95
Stump Blasting........ 25
Stump Blasting: Advantages of........ 14
" " Cost of........ 16
" " Gauging the Charge........ 15
" " Proper Explosive for........ 15
Subsoil Blasting........ 69
Swamp Draining........ 60

Tamping........ 103
Thawing of Dynamite........ 97
Thawing Kettles........ 97
Transportation of Explosives........ 97
Tree Felling........ 41
Tree Planting and Cultivating........ 79

Well Sinking........ 64
Western Cedar Stumps........ 31
Western Fir Stumps........ 31
Western Pine Stumps........ 31
Wire: Leading and Connecting........ 110

IF

After reading this book you want further information, write us or our nearest

Branch Office

(See list on last page)

E. I. du Pont de Nemours Powder Co.

ESTABLISHED 1802

Wilmington, Del., U. S. A.

BLASTING HARDPAN IN KANSAS

INTRODUCTION

NOT so very many years ago the farm was about the last place where one would expect to find up-to-date mechanical appliances. Steam, explosives, electricity—the factors which have made the world in general, and America in particular, what it is today—were indispensable for the railroad and the steamboat, the factory and the mine, but the farmer's work was generally run on a comparatively small scale, and was mostly done by the muscles of men or draft animals. Today, however, the farmer as well as the manufacturer, the contractor and the miner, fully appreciate the value of labor-saving devices. Every up-to-date farm, large or small, has modern machinery that tends to reduce the cost of production or to improve farm products.

Explosives were first used in warfare and hunting, then for blasting. Their early use in mining and excavating was very limited, but it did not take long to find out that explosives, if used in the right way, could do in a moment as much work as a man, or even a machine, could do in many days. As soon as this was understood, explosives were applied to new kinds of work and special kinds of explosives were produced, particularly adapted to these new uses.

The history of the use of explosives in farming has been much the same as in other industries. At first it was supposed that they could only be used for splitting stumps, so that they could be dug out more easily. Later on, after stump blasting had been carefully studied, and more suitable and cheaper explosives manufactured, it was found that by far the cheapest and quickest way to get rid of a stump of any size, no matter what kind of soil it stood in, was to lift it completely out of the ground, and split it up at the same time, with explosives.

FARMERS' INSTITUTE IN VIRGINIA ATTENDING DU PONT STUMP BLASTING DEMONSTRATION

About this time it was discovered that it was cheaper to blast out boulders and plant the ground they occupied than to plow around them. Then some enterprising farmer who had had trouble with foundations settling, decided that it was a good plan to build his house and his barns on rock and learned that it did not cost much to do the necessary blasting for foundations and cellars. This led to the idea of blasting the rock encountered in digging drains, and in using a small quantity of explosives to hurry along the work of digging holes for fence posts and for poles. The rock met with in sinking wells was soon cut through with explosives. Early in the spring when the ice came down the streams, and a gorge at the bridge threatened to cause a flood and carry away the bridge too, a little dynamite quickly relieved the situation. In lumber districts, log jams were started in the same way, and the use of a small quantity of explosives saved time and trouble on the rollway.

Within the last few years dynamite has been used for blasting subsoil and ditches, and it is doubtful if its use in any other kind of work, represents so great a saving, or has so many advantages.

A few years ago the plan of breaking up hardpan and other impervious subsoils was tried by Samuel J. Crawford, then Governor of Kansas, and a number of other influential Kansans who were interested in farming. The results of their experiments were so successful that hardpan is now being blasted quite generally, and many acres of land which were practically worthless before they were blasted, are now bearing phenomenal crops.

The most recently discovered method of saving money by using dynamite is in blasting ditches. Dynamite has long been used to shatter rock encountered in digging drains and ditches, but ditches through earth have never been excavated entirely with explosives until recently.

Many fruit growers know how great a help dynamite is in planting fruit trees and in keeping them thrifty. The ground where the tree is to be planted can be loosened up, and the hole partly dug, in a moment, by exploding a very small charge of dynamite a short distance below the surface.

Good roads about the farm are practically indispensable, and they cannot be built quickly or cheaply without at least some explosives.

Explosives are also used to advantage for splitting logs for rails or any other purpose, for felling trees, destroying old buildings and for almost any kind of work where a strong force quickly applied, is desirable. Dynamite is simply concentrated power or condensed strength. In order to have it work in the correct way, and move in the right direction, it must be properly harnessed and the bit kept in its mouth, but this is easily done if the directions given in this handbook are followed intelligently.

As time passes, farmers will find other ways of saving money by using explosives, for the demand for farm products is increasing so rapidly that greater acreage must be put under cultivation, and the old acres must increase their yield. In order to do this successfully and to meet competition, every device and arrangement that really saves labor and makes for economy, either in the present or the future, must be adopted. The farmer who fails to realize this, and still believes that he can get along in the old way, will soon find himself hopelessly defeated.

It has always been the policy of the Du Pont Company to study thoroughly the various kinds of work in which explosives can be used to advantage, and then to produce explosives best suited for the work to be done. In addition to this the company has adopted every possible device which will help it to manufacture explosives at a low cost, and at the same time improve their quality. It has a large number of chemists constantly engaged in improving its explosives, and in testing the daily output of the various works, so as to keep its products up to standard. The company also has a force of men experienced in the use of explosives, who go about the country studying the work in which explosives are employed, and instructing consumers how to use them to the best advantage. All of this is expensive, but it enables the Du Pont Company to be **sure** that it is selling **the very best explosives that can be made,** and also to furnish the consumer with the kind of explosive which is most suitable for the work to be done.

E. I. du Pont de Nemours Powder Co.

December, 1910

THE FIRST STAGE

CLEARING THE LAND

BLASTING OUT STUMPS

FELLING TREES

SPLITTING LOGS

BLASTING BOULDERS

A FUTURE SUBJECT

CLEARING THE LAND

EVER since the first white man started a permanent settlement in this country, the transformation of forests into farms has been carried on continuously, and at an ever-increasing speed. So great is the total of land clearing operations today, and so large is the annual expenditure in this work, that it has become a factor of prime importance in the national economy, and the necessity for a careful study of all of its phases, to the end of improving the various methods now in practice, is everywhere recognized.

If, after the passing of the wood-chopper and the sawmill, the land were ready for the plow, there would be no problem to solve; but a great area covered with stumps—thirty, fifty, a hundred, two hundred to the acre—often with boulders scattered here and there, is very far from a crop-bearing proposition. This, however, is what very many of our farms are made from, and the question is, how to bring about the necessary transformation without spending more than the probable returns will warrant.

If, for example, uncleared land costs twenty-five dollars an acre, and after being cleared will only yield a fair return on the outlay of fifty dollars an acre, it is evident that it must be cleared for not more than twenty-five dollars an acre. The problem seems a simple one, but its solution is sometimes difficult—we know what is to be done, the question is, how to do it?

For generations farmers have chopped, burned and dug away at stumps which have slowly yielded to their efforts. Today a quicker and more effective way of getting rid of them is imperative. The subject has been studied and experimented with for years by the Agricultural Departments of the National Government and of different states, by many large land companies, by railroad corporations and by thousands of individuals. Numerous ideas and plans have been suggested and worked out, but none of them has been successful or economical without the use of dynamite. When explosives are properly used, stumps and boulders are not only blasted out of the ground but

A FIVE-FOOT DIAMETER WASHINGTON FIR WINDFALL SHOWING THE ENORMOUS SPREAD OF THE ROOTS

are at the same time broken into pieces which can be easily handled—burned, if stumps, or, if boulders, used for building roads, etc.

When a stump is properly blasted, but little of the soil is thrown out with it, and it is not necessary to spend time and money clearing the earth from its roots, and filling a great hole in the ground, as must be done when the stump is taken out with a stump puller. Neither is any special fertilization of the spot formerly occupied by the stump, necessary to bring it into proper crop-bearing condition, as must be done after a stump has been burned out.

Only a very small outlay is necessary at one time when explosives are used, and one man can work with them just as well as a dozen or a hundred. When the proper explosive is selected, and when it is used in the right way, there is no cheaper method of getting rid of stumps or boulders.

There are many kinds of dynamite, each expressly intended to do a particular kind of work; and as the conditions under which stumps and boulders are blasted differ widely, it is not possible to recommend any one grade for general use.

On the Pacific Slope, where the stumps are of great size, Hercules Stumping L. F. Powder is the favorite, but in other parts of this country, where the stumps are smaller, Red Cross Dynamite is used almost exclusively. The popularity of these two brands is probably due to the fact, that they are the only ones manufactured that are similar in form, appearance, and action to the standard dynamites, but which do not freeze at temperatures of 45° or 50° F. as the other brands do. Hercules Stumping L. F. Powder and Red Cross Dynamite do not freeze until water freezes, frequently not even then, and will thaw out in a very short time when they are exposed to a temperature that will melt ice.

It is equally difficult to state accurately, the quantity or strength of the explosive necessary to blast out a stump of a given diameter. This is because the size of the stump is not the only factor, and perhaps not even the most important one to consider, when estimating the charge required. Whether the soil in which the stump stands is wet or dry, light or heavy, the kind of wood, age of the stump, the nature and position of the roots, etc., are all matters of great importance when determining the quantity, strength and location of the charge of explosives. Careful records have, however, been kept of the cost of explosives, including blasting caps and fuse, or electric fuzes, used in blasting stumps in different parts of the country, some of which we give below:

	Average Diameter	Average Cost of Explosives per Stump
768 Southern Pine Stumps..................	29"	30c
78 Oak, Walnut, Gum, etc., Stumps in Illinois .	30"	53c
329 White Pine, Oak, Maple, Birch, etc., Stumps in Michigan.......................	32"	47c
37 Apple, Ash, Oak and Chestnut Stumps in Pennsylvania......................	34½"	56c
77 Washington Fir Stumps.................	50"	$1.13

These figures were taken from records of the Du Pont Powder Company. The explosive used in most cases, except for blasting the Washington firs, was Red Cross Extra (40%) Dynamite. The fir

WASHINGTON FIR STUMPS

stumps were blasted with Hercules (20%) Stumping L. F. Powder and Judson Powder R. R. P. (5% strength). These calculations are based on retail, not wholesale prices.

Records kept by A. J. McGuire, Superintendent of the Northeast Experimental Farm of the University of Minnesota, show even lower costs.

Some of Mr. McGuire's records are as follows:

	Average Diameter	Average Cost of Explosives per Stump
255 Popple	14"	12c
255 Jack Pine, Norway Pine and White Pine	14 1/3"	18c
395 Birch, Ash, Spruce, Pine, etc.	20"	16c

Mr. McGuire used 25% to 40% Red Cross Ammonia Dynamite, and states that the best and most economical results were had with 25% and 27% grades.

The Iowa State College recently blasted eighty-two oak and elm stumps and trees averaging 20 inches in diameter, at an average cost of about 38 cents each for explosives.

On a large land clearing operation in Minnesota during the summer of 1909, eight thousand, nine hundred and seventy stumps were blasted out. Although a considerable number of these were large pine stumps, an average of less than three-quarters of a pound of dynamite per stump was used.

Accurate records of the cost of blasting stumps on a Long Island farm, including the wages of the men who did the work, were recently kept by representatives of the Long Island Railroad Company. The entire cost of blasting out and burning up one hundred stumps was only $16.00.

Although one unaccustomed to using explosives might find the costs running higher at the start than some of those given above, it does not require an unusual amount of experience to learn approximately the minimum charge required to blast stumps.

When cut-over land, which is covered with stumps and boulders, can be cleared, and turned into farms at a profit, it is hard to under-

DU PONT REPRESENTATIVES DEMONSTRATING STUMP BLASTING IN NEW HAMPSHIRE stand why anyone should let stumps or boulders take up valuable land and then plow around them year after year. A lot of time is wasted swinging around even a few stumps and boulders when plowing a field, to say nothing of the damage to plow, harness and team if a root is struck. Besides this, each medium-sized stump, with its spreading roots, or even a comparatively small boulder, will take up the space of many stalks of corn or of other crops. It has been estimated that a single stump of average size occupies as much ground as would support six hills of corn.

The corn from six hills would in a very few years pay for enough dynamite to blow out many stumps and boulders.

One suggestion which we would make to those who have land to clear is: Always do your stump blasting, if possible, when the ground is wet. Almost every kind of ground when it is wet, offers stronger resistance to the action of dynamite than it does when dry. Therefore, when the ground is wet a stump or boulder can be blasted out with less dynamite than when the same ground is dry.

One of the most objectionable methods of trying to get rid of stumps is burning them out. When stumps have been blasted out and split up with dynamite, it is an easy matter to heap up the pieces and burn them, but to burn a standing stump is a different proposition. Those who have tried it can testify to the time spent in keeping the fire going, and that it is practically impossible in this way to get rid of much of the stump below the surface of the ground. Probably the worst feature of burning out stumps is the damage done to the ground by the fire, which burns out the humus to such an extent, that it requires much cultivation to bring the ground where the stump was burned into good condition. The following from the Tacoma, Wash., "Ledger" of October 20, 1909, explains very clearly the damage done to new land by burning out stumps:

DU PONT STUMP BLASTING DEMONSTRATORS AT VIRGINIA FARMERS' INSTITUTE, 1910

> "Last summer Prof. W. J. McGee of the Department of Agriculture, in co-operation with Prof. Henry Landes, Dr. Benson and Dr. Fry of the State University, studied the best methods of utilizing cut-over or logged-off lands. Prof. McGee gives some of the findings, and they argue strongly against burning cut-over lands. Fire destroys the humus in the soil, and when the humus is gone, the fertility is greatly lessened. The danger of burning logged-off lands is that the fire will burn the humus for some distance below the surface. When the humus is burned out, it takes time to build up a fertile soil again. The fire destroys the work it took nature many years to do."

The implements needed in stump blasting are few and inexpensive, and most of them are always to be found on the farm.

Big stumps like those found along the Pacific Coast usually require a comparatively bulky charge of low-grade explosives, which means that a large hole must be dug. In case the stump is very large it is best to dig a trench under the stump, using a little dynamite to assist in the digging. In this work a pick or mattock and a long-handle shovel are necessary. As a matter of fact these two tools, and an axe, are usually required, no matter how the hole for the explosive is made.

For medium-size and smaller stumps, a two-inch wood auger to drill under the stump (and into the tap root if there is one), is necessary, and a crowbar with one pointed end and one chisel end is very useful. One of the most important implements used in stump blasting, is the tamping stick. This must have no metal about it. The tamping sticks used on the Pacific Coast are six to eight feet long and two to three inches in diameter. For tamping the holes under other stumps a stick five or six feet long and one and a half inches in diameter is large enough.

For blasting large boulders it may be necessary to use a hammer and hand drill, unless they are to be broken by mudcapping or by placing the explosive underneath them. The hole under the boulder can be made with a crowbar. If the boulder is drilled, a small tamping stick, from one-half to three-quarters of an inch in diameter should be

A WELL CLEARED FIELD—STUMPS SO SPLIT THAT THEY WILL BURN EASILY used. If the charge is put under the boulder, the tamping stick should be one and a half inches in diameter and five or six feet long. Hardwood tamping sticks are of course best, because they last longest.

ONE MORE OUT OF THE WAY

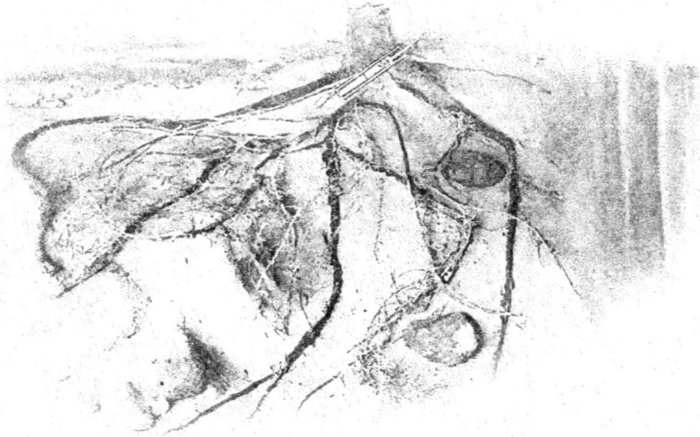

SECTIONAL VIEW OF STUMP
SHOWING RELATIVE POSITION OF ROOTS AND EXPLOSIVE CHARGE

BLASTING STUMPS

IT IS usually necessary in blasting stumps to place the charge as close as possible to the toughest part of the stump, so that the part most difficult to break will be hit first and hardest. Generally this spot will be directly under the middle of the stump, and it is sometimes necessary to bore into the tap root. There are cases where a very big stump is rotten at the middle, but has several large, branching roots. In such instances better results will be had, if the charge is increased a little and located deeper in the ground, or if under each of the large roots a smaller charge is placed, and all of them fired simultaneously with a blasting machine.

In order to keep the explosive from splitting the stump, and wasting a part of the force which should be expended in lifting it out, it is the practice of some blasters to wind a stout chain around the stump several times. It is a good idea to use a chain with all hollow stumps.

STUMP BOUND WITH CHAIN

DU PONT REPRESENTATIVES DEMONSTRATING STUMP BLASTING IN GEORGIA

Some large stumps can be effectively removed by exploding simultaneously, charges loaded in holes bored from different sides, so that they will intersect under the center of the stump. The cartridge containing the detonator (blasting cap or electric fuze) should be loaded first, so that it will be at the intersection of the holes; then the different holes should be loaded, making sure that the first cartridge in each hole touches the primer (the cartridge containing the detonator). When the primer is exploded, it will explode the entire charge.

If the above directions are closely adhered to, the general run of stumps can be blown out satisfactorily and cheaply. The stumps of some kinds of trees, however, require special treatment to get the best results, and detailed instructions concerning them are given in the following pages.

Southern Pine Stumps

The use of explosives in blasting southern pine stumps, especially in very sandy soils, has not been generally recommended in the past, because the cost has seemed high in comparison with the value of the land. The increased value of land and the improvements in explosives have, however, now made it entirely feasible to remove these stumps with dynamite. This work can be done with less dynamite, if a good team of horses with a bull chain is used to pull out pieces and drag them from the field.

Pine stumps should be considered in two classes—those with tap roots and those without. The first part of a pine stump to rot is the tap root, and a field that looks very discouraging will often be found easy to clear.

In blasting southern pine stumps the important feature is to place the explosive close to the point of greatest resistance, which is in the tap root directly under the center of the stump. This tap root can usually be located by digging underneath the smooth side of the stump. When the tap root is located, dig an opening, one to two feet deep, down by the side of the tap root; then bore an auger hole two inches

BLASTING STUMPS IN VIRGINIA

in diameter into the tap root on an angle of from 35° to 50°. The auger hole should be bored not less than three-quarters of the way through the tap root; care being taken not to drill entirely through, for by so doing a great deal of the explosive force would be wasted in the back of the hole, which would then be in the ground. Load with from one to three 1¼ x 8-inch cartridges (depending on the size and age of the stump), of Red Cross Extra (40%) Dynamite, and tamp firmly to the surface with damp clay. The blast will cut off the tap root twenty to thirty-six inches below the surface and turn out the stump in pieces. These stumps can also be blown out without taking the time to bore into the tap root, if a little larger charge of dynamite be placed firmly against the tap root a foot or two below the surface and closely confined with tamping. A convenient implement for making the hole in the ground when blasting in this way is a three-inch post-hole auger.

When the tap root is to be bored into, it is advisable to use a two-inch wood auger, as the dynamite can then be more closely compressed under the center of the tree or stump where the resistance is greatest. If clay cannot be found to tamp the charge under the first stump blasted, use sand. After one stump is blasted, you can usually find good tamping material at the bottom of its roots.

Careful record was recently kept by one of our representatives, in order to arrive definitely at the exact cost of explosives necessary to properly blast out these southern pine stumps. Three hundred and twenty-five stumps were blasted which averaged in diameter 28½ inches, and the cost of explosives including dynamite, fuse and blasting caps, or electric fuzes, averaged a little more than 18 cents per stump.

Southern pine stumps without tap roots are sometimes found in land having a sandy top soil with a hard subsoil. In this case Red Cross S. P. 1 (30%) Dynamite or Judson Powder F. F. F. (20%) may give the best results. The charge should be placed under the middle of, and as close to the stump as possible.

Washington Fir Seven Feet in Diameter

Blown out with 25 pounds of Hercules Stumping L. F. Powder

BEFORE THE BLAST

AFTER THE BLAST

Second-Growth Stumps

There is often directly under a second-growth stump, the decayed remains of the original stump; this is soft, and the force of the explosive when placed on it, seems to merely scatter this dead wood and has no marked effect upon the stump. To overcome this difficulty, it is a good plan to dig under the stump and place a good-sized flat stone between the roots, leaving only room on top of the stone for the dynamite. Damp clay should then be firmly packed around the dynamite. This gives the explosive sufficient resistance to lift out the stump. Red Cross S. P. 1 (30%) or Red Cross Extra (40%) Dynamite should be used. It should be remembered that best results will be had from the explosives recommended for blasting the above stumps, if they are exploded with No. 6 (red label) or stronger detonators.

SHOWING HOW THICK THEY GROW

Western Fir, Pine and Cedar Stumps

In the states of Washington, Oregon and California, where the rainfall is large and the ground in the forests is always damp, many of the trees grow to great size—some being eight or ten feet in diameter. The roots of these trees usually spread out near the surface and do not grow deep into the ground, as might be expected, tap roots being extremely rare. The object when blasting the stumps of these trees

A CALIFORNIA REDWOOD STUMP

BEFORE THE BLAST

The circumference of this stump at the ground line was 75 feet. It was a *single* stump in the ground, but divided into two just above the surface.

is not to split them, but to bring them out entire at one blast, with all of the roots possible, because if the charge of explosives is so gauged and located as to split the stump, it generally fails to bring out all of the pieces. As the principal object is to get out as much of the stump as possible at a minimum cost, it is better to blast it out first and then it can be easily split afterwards, by means of a small quantity of dynamite exploded in auger holes.

The common rule in blasting these stumps is to use one and one-half pounds of Hercules Stumping L. F. Powder per foot of diameter, with stumps up to four feet, when the bottom is clay. For larger sizes it is advisable to use two to two and one-half pounds of this

OREGON PINE STUMP—PREPARING FOR THE BLAST

powder for each foot in diameter. For stumps in gravelly or loose ground one pound more should be used for each foot in diameter.

The charge of explosives is best placed when there is sixteen to twenty-four inches of earth between the charge and the bottom of the stump. This results in the force of the explosion radiating to all sides, lifting the stump clear of the ground, and bringing with it the greatest length of roots. If the charge is placed too close to the stump, the effect is to split it, leaving the roots to be dug out at extra labor and expense. It is advisable with large stumps to chamber, or expand,

A CALIFORNIA REDWOOD STUMP—THE BLAST

NOTE.—The main part of the stump is split into several pieces and lifted straight up, while the smaller sections, which were blown out first, are already falling to the ground. This shows that no more explosive than necessary was used in this blast.

the bottom of the hole, so that the entire charge can be concentrated under the center of the stump. To do this a hole is bored beneath the stump by means of a dirt auger or small post-hole auger. In this hole a small charge is usually exploded with fuse and blasting cap and no tamping whatever is used. The quantity of explosives to be used in chambering, depends of course on the size of the charge necessary to blast out the stump. Usually from half a cartridge to two cartridges of Hercules Stumping L. F. Powder will be sufficient. After the hole is chambered the blasting charge should be loaded in the usual way and thoroughly tamped.

Redwood and *Bigtree Stumps

The best explosive for these stumps is Judson Powder R. R. P. It is a comparatively slow-acting explosive, and has more of a lifting and heaving, than a shattering effect. It is granular, and is packed in twelve and one-half pound paper bags which are enclosed in wooden cases similar to those in which regular dynamite is packed.

The way to approximately estimate the quantity of Judson Powder R. R. P. necessary to blast out stumps larger than eight feet in diameter, is to square the largest diameter in feet, the result being approximately the number of pounds required. For example, if a stump is eight feet in diameter the largest way, the charge of Judson Powder R. R. P. should be about sixty-four pounds. Stumps less than eight feet in diameter require a little greater charge for their size than do the larger stumps, and the rule with them is to use as many pounds of Judson Powder R. R. P. as eight times the largest diameter in feet. On this basis a stump six feet in diameter would need about forty-eight pounds of powder. However the successful blasting of these large stumps depends largely on the judgment of the blaster, and these rules can only be considered as a general guide. This can easily be understood when it is remembered that, owing to difference in soil or some peculiarity in the growth of the tree, it sometimes requires the

*"Bigtree" is the name given to the "Sequoia Washingtoniana," one of the gigantic trees of the Pacific Northwest.

A CALIFORNIA REDWOOD STUMP

AFTER THE BLAST

The tree had been cut from this stump for twenty-five or thirty years, but the stump was perfectly solid. It made about thirty-five cords of wood.

same quantity of explosives to properly bring out a stump six feet in diameter as it does one eight feet in diameter.

In blasting these stumps a trench is dug large enough to permit placing the entire charge of explosives directly underneath the center of the stump. A little dynamite blasted in holes punched with a crowbar will prove of great assistance in digging this trench.

If the ground is wet, the charge should be placed in a waterproof bag, as Judson Powder R. R. P. is not waterproof and is quickly damaged by contact with water.

Judson Powder R. R. P. can be properly exploded only with a primer of 40% (or stronger) dynamite. The sizes of the primers required for different charges of Judson Powder R. R. P. are as follows:

Charge of Judson Powder Pounds	Primer of Dynamite Number of 1¼ x 8-Inch Cartridges
10	1
20	2
50	4
300	25

A detonator equivalent to, or stronger than a DU PONT No. 6 (red label) Blasting Cap, or Victor No. 6 (red label) Electric Fuze, should always be used in one of the priming cartridges. The proper way to make the primer is shown on pages 100 and 102. If several cartridges are used as a primer, they should be tied in a compact bundle with the primed cartridge in the center. If blasting cap and fuse are used in the priming cartridge, care should be taken in placing the primer to prevent any contact between the fuse and the Judson Powder R. R. P. as the latter is very inflammable. The charge should be firmly tamped.

Avoid being on the same side of the stump as the trench when the blast is fired, as fragments, etc., are thrown with more violence and to greater distances on that side.

Hercules Stumping L. F. Powder is also used in blasting California redwood stumps. The illustration on page 32 shows two large redwood stumps which had practically one root below the surface,

VANCOUVER, WASH., STREET CARS ADVERTISING DEMONSTRATIONS OF STUMP BLASTING WITH DU PONT DYNAMITE although this root had separated above ground into two trees. The circumference of the stump just above the surface of the ground was seventy-five feet. This stump was completely removed, as shown on pages 32, 34 and 36, with ninety-three pounds of Hercules Stumping L. F. Powder. Six trenches were dug under the stump at different points, five of these being loaded each with twenty-five 1½ x 8-inch cartridges of this explosive, and the sixth with thirty 1½ x 8-inch cartridges. These charges were then connected up electrically, and the trenches were thoroughly and compactly tamped above the explosives to the surface of the ground. The six charges were then fired simultaneously with a blasting machine. The illustration on page 32 shows the blasting machine used and the cartridges of Hercules Stumping L. F. Powder on the ground preparatory to charging the trenches.

This stump had stood from twenty-five to thirty years, but was perfectly solid. It made about thirty-five cords of wood after it was blasted.

Cypress Stumps

Cypress stumps are found, as a rule, in swamps where the soil is a soggy muck often covered with water. Through the land-reclaiming operations in the Southern States many of these swamps have been drained, leaving land of wonderful fertility. The cypress stumps have no tap root, but have large "spreaders" reaching out in all directions to such an extent that they are interwoven with those of neighboring stumps, forming a tangle of roots that never rot. The strongest and quickest dynamite in the hands of careful blasters gives the best results in this work. The common practice is to place 1¼ x 8-inch cartridges under each of the principal spreaders, and fire all simultaneously by means of a blasting machine. The cypress wood, being extremely soft, splits easily, and the quick explosive shatters the stump and releases it from the entangling roots.

In order to obtain the best results in blasting cypress stumps the electric system of blasting should be employed. The Du Pont Company recommends the use of 60 per cent. Semi-Gelatin, detonated with a No. 6 (red label) Electric Fuze for the work.

When cypress stumps are not blasted until after the swamps have been drained off, Red Cross Extra (40%) Dynamite should be used.

FELLING A TREE—THE BLAST

FELLING TREES

Occasionally when clearing land of growing timber, it is of advantage to blast out the entire tree and saw off the root afterwards. The process here is exactly the same as in stump blasting, but little if any more dynamite being required to bring out the tree, roots and all, than to blast the stump after the tree has been cut. The blast lifts the tree straight up a foot or two; then it falls, generally with the wind.

SPLITTING STUMPS AND LOGS

When stumps, particularly large ones, are blasted out whole or nearly so, it is usually necessary to split them up so that they can be conveniently handled or burned. This can be readily accomplished with dynamite; only a small quantity of explosives being required, if the charge is properly tamped in auger holes bored part way through the stump.

In the South the pine stumps are very large producers of turpentine and by-products. Before the wood in the stump can be distilled, it must be broken into pieces small enough to suit the retort. Nothing is so effective as dynamite for breaking up a stump for this purpose. Charges of a few inches of Red Cross Dynamite exploded simultaneously in several auger holes bored in the stump, will shatter it up into exactly the size required.

When logs are split up to be burned quickly, the same method is used as when splitting stumps; but if they are to be split for fence rails, cord-wood, charcoal, or other purposes where comparatively even and regular sections are required, blasting powder or Judson Powder R. R. P. should be used. These explosives are so much slower in action than dynamite that a series of properly gauged and properly placed charges will split a log along the grain, just as evenly as if a number of wedges were used. This method of splitting logs is so much quicker, cheaper and easier than any other, that those who have once become proficient at it, never give it up. Auger holes are bored along the line of the grain, about one-quarter to one-half of the way through the log, the depth of the holes and the distance between them depending on the kind of wood, the grain, and the diameter of the log. A few

FELLING A TREE
The result—Tree laid down in proper place without injuring adjoining fence

ounces of FF blasting powder is put into the bottom of each hole, care being first taken to see that the hole is dry, then wooden plugs are driven firmly into the tops of the holes to tamp or confine the charge. In some kinds of wood it is best to leave a considerable air space between the bottom of the plug and the powder. The plugs must have a groove in the side large enough to admit the electric squib wires or fuse. As blasting powder is exploded by a spark or flame it is not necessary to use a detonator (blasting cap or electric fuze) when blasting with it. If electric squibs and a blasting machine are used for exploding the charges, they can all be fired simultaneously. This usually is the best and cheapest way, as a little less powder is required than when the charges are exploded separately with fuse. Electric squibs are similar in appearance to electric fuzes, except that they have a paper capsule instead of a copper cap. They do not explode when the electric current passes through them, but ignite the blasting powder by a flash. When using electric squibs, it is only necessary to have the groove or channel in the sides of the wooden plugs large enough for the two small wires to run through it, if the cap of the electric squib is put in place before the plug is driven in. When driving the plug care must be taken that the wires are kept free, and that the insulation on them is not abraded. If it is not convenient to provide wooden plugs in this work, damp clay tamping may be used on top of a wad of newspaper. A log two feet in diameter, and four or five feet long, can usually be split in two with one two-ounce charge of FF blasting powder fired as described above. Longer logs require two or more holes, and logs of greater diameter require heavier charges. The holes should be from one and one-eighth to two inches in diameter.

Logs up to six feet in length can be split at once into a number of pieces, by exploding a single charge of Judson Powder R. R. P. in a hole bored about twelve or fourteen inches straight into the center of one end. Two ounces of Judson Powder R. R. P. is about the right charge for a log two feet in diameter, but about an ounce and a half of 40% dynamite and a No. 6 detonator are required to explode the Judson Powder R. R. P.

BLASTING BOULDERS IN VERMONT

BOULDER BLASTING

THE BOULDER—BEFORE THE BLAST

THERE are three ways in which boulders can be blasted. These are known as "Mudcapping," "Snakeholing" and "Blockholing." "Mudcapping" and "Snakeholing" are the easier and quicker methods, but require more dynamite. It is almost impossible to shatter large round boulders of hard rock by either of these methods, without using an excessive quantity of explosives, which makes blockholing necessary with boulders of this kind.

Mudcapping ("Doby Shooting" or "Blistering")

Boulders may frequently be broken by exploding a small charge of dynamite on their surface. The charge should be placed on the spot which would be struck with a sledge if the boulder was to be broken in that way. The dynamite should be packed in a solid mass by slitting the paper cartridge shells, but should not be spread over the surface of the boulder any more than absolutely necessary. A blasting cap crimped on to fuse should be placed in the middle of the charge, and the whole covered with six inches of damp clay or sand. This material should be pressed firmly over the mass of dynamite,

BLASTING A BOULDER

care being taken not to cover the end of the fuse. It is advisable if the boulder is deeply imbedded in the ground, to dig away or loosen some of the earth surrounding it before blasting.

If the boulder is cracked or seamy, the best results may be secured by placing the charge in some depression and covering it with a quantity of clay or sand. This will furnish more resistance and make available a greater force from the explosive.

The quantity and grade of explosives required naturally depend on the size and shape of the boulder. The "grain" and kind of rock are also important points. Red Cross 40% Dynamite is the grade recommended, and the quantity required will vary from one to ten pounds, the latter quantity being necessary only when the boulder is a very large one. The dynamite should be exploded with a No. 6 (red label) detonator.

Snakeholing.—Proceed as in stump blasting, taking care that the hole in the ground be made under a flat or hollow side of the boulder and not under a bulging side. Make a hole with a crowbar or a dirt auger, close up under the center of the boulder and load the dynamite into the hole in the same manner as you would for stump blasting. Care should also be taken when the charge is placed, to leave no means by which the force of the dynamite may escape. If it has not been thoroughly tamped, or if it is too near the surface of the ground, and not in the proper position beneath the boulder, the dynamite may blow the dirt out and leave the boulder untouched.

Red Cross Dynamite of 30% to 40% strength should be used when breaking up boulders in this way. Only from one-half to two-thirds the quantity of dynamite, that would be required to mudcap the same boulder, is needed provided the boulder has a hollow or concave side underneath. The results are better in damp, heavy soil than in light or sandy soil.

Blockholing.—This is the most economical method of using dynamite to break up boulders, and although it takes some time and labor to drill the one or more necessary holes in the boulder, it will often be found the most satisfactory in the long run. The holes in large boulders should be an inch or more in diameter, while three-quarters or seven-eighths of an inch will answer for the smaller ones. To properly break up a boulder weighing about ten tons, a hole should be drilled in it from ten to twenty inches deep, as the shape and grain of the rock may demand, and large enough in diameter to hold a pound or more of dynamite with the necessary tamping. Smaller boulders require holes from four to six inches in depth, which, if necessary, can be filled full of dynamite, and no tamping used.

Red Cross S. P. 1 (30%) Dynamite will usually give best results in blockholing, unless the boulders are very large and hard, when Red Cross Extra (40%) should be used.

THE BOULDER—AFTER THE BLAST

THE SECOND STAGE

GETTING THE FARM IN SHAPE

DIGGING DITCHES

DRAINING SWAMPS

GRADING ROADS

EXCAVATING FOR BUILDING FOUNDATIONS AND CELLARS

SINKING WELLS

DIGGING HOLES FOR FENCE-POSTS

PUNCHING THE HOLES FOR DITCH BLASTING

GETTING THE FARM IN SHAPE

WHEN the land has been cleared the work of getting the farm into shape begins. Ditches must be dug to properly drain the fields—particularly the swampy ones, where often the best soil is found. Roads must be constructed, wells sunk, barns, houses and other buildings erected, fences built, and so on. Dynamite is necessary in every one of these operations, if the work is to be done expeditiously and economically. The information to be found in the succeeding pages, as well as that already given, has been secured entirely from farmers and others who have used explosives for the various purposes described. Most of this information has also been confirmed by tests and experiments conducted by one or more of the numerous blasters in the service of the Du Pont Company.

DITCHING

Ditches, long or short, deep or shallow, are necessary on every farm, and should always be dug with explosives whether they go through rock, shale or only earth. The ditching which pays best is that which drains swampy land. There are about seventy million acres of this land in the United States, much of which would bear magnificent crops, if drained. The cost of draining by the usual method of digging ditches by hand or machine, has been almost prohibitive, with the result that this land is still lying idle. Now that it has been found that dynamite will dig ditches through any swamp where a man can penetrate, no matter how thick the vegetation covering it may be, there is no reason why the large majority of this seventy million acres of swamp land should not be cultivated.

BLASTING A DITCH

Dynamite is just as effective for digging ditches and channels in wet, swampy ground, even though it be covered by several inches or a foot of water, as it is in dry ground. A channel or ditch dug with dynamite, and particularly one cut through dry or sandy ground, is not so even and regular at first as one dug by hand or machine, but will square up very soon after water fills it and, as the banks have a good slope, there is little caving afterward.

When a ditch is blasted there is no outlay for expensive equipment, because the only machinery necessary is an iron bar, pointed at one end. In hard, dry ground, a sledge or maul is needed to drive the bar down to the necessary depth. There is no delay and expense getting machinery into place through swamps and thickets. Not even a team is needed when ditches are dug with dynamite, for one or two men can easily carry sufficient dynamite to dig a ditch four or five hundred feet long, four or five feet wide, and three or four feet deep.

When ditches are dug with dynamite the material taken from the ditch is practically all thrown out by the blast, and little or no shoveling is necessary. This material is not heaped up along the sides of the ditch, where it would occupy land that should be raising crops, and would prevent free drainage into the ditch, but is spread evenly over the ground for a considerable distance on each side of the ditch.

Ditches can be dug with dynamite at any time on a couple of hours' notice, as dealers who handle explosives can be found in almost every locality. In wet weather, especially after heavy rain falls, when it may be necessary to cut ditches and channels in a hurry, this method is invaluable.

It is just as easy to dig a curved ditch with dynamite as it is to dig a straight one, because the center of the ditch follows the line of holes in which the dynamite is loaded.

How It Is Done

Dynamite digs the cleanest and most regular ditches in wet clay or gumbo. In this kind of ground it is not necessary to put the holes so close together, or to use quite so much dynamite as when the clay is only damp. In dry clay, sand marl, or other loose ground, the weaker and consequently less expensive grades of dynamite, give best results.

A DITCH BLASTED WITH DYNAMITE—WILL SQUARE UP AS SOON AS FILLED WITH WATER

If the soil is very light, it may occasionally be necessary to "trim up" the ditch a little by hand after the blast, but even then it is at least twenty-five per cent. cheaper to dig the ditch with dynamite than by machine, and it is also many times quicker.

When ditches are dug in wet clay it is best to explode the dynamite with blasting cap and waterproof fuse. If the holes are spaced the proper distance, only the hole in the middle of the row requires a blasting cap, as the explosion of the dynamite in this hole explodes that in the next holes on either side, and so on almost instantaneously from hole to hole, to the two opposite ends of the row. This plan of exploding the dynamite in one hole by that in the next hole, instead of putting a detonator in each hole, works best in wet ground, and when the ground and water are not cold. The temperature of the air, water and ground is an important point. If it is below 50° F., the dynamite may become so insensitive that the charge in one hole will not explode the next one. It is, therefore, recommended that whenever possible, this work be done when the ground is wet and the weather warm.

When all conditions are favorable the holes can be spaced two feet apart, put down three feet and loaded with one cartridge, 1¼ x 8 inches of No. 2-SS (50%) or No. 1 (60%) Hercules Dynamite or of B (50%) or B+ (60%) Atlas Dynamite, for a ditch four feet deep and six feet wide at the top. Three men can dig a thousand feet of this ditch in three days, with two hundred and fifty pounds of dynamite, and at about one-third of the cost of any other method. Fifty, sixty, or even a greater number of holes, can be blasted simultaneously with a blasting cap in the middle hole only. Sometimes as much as a quarter of a mile of ditch is blasted in this way at once, but unless a number of men are loading the holes it is not advisable to blast so much at one time, because the dynamite in the holes first loaded will be under water so long, before the last holes are loaded, that it may deteriorate. The middle hole should be loaded heavier than the others—say with three cartridges instead of one cartridge, one of them being the "primer"; that is, the cartridge in which the blasting cap crimped to the fuse has been placed. It is also a good plan to load the next holes on either side of the middle one with three cartridges instead of one. When water covers the holes, or almost fills them, it acts as a tamping, and no other tamping is necessary. The middle hole should be loaded last, so as not to keep the blasting cap and fuse under water longer than absolutely necessary. The blasting cap **should be carefully crimped to the fuse with a cap crimper,** and **not** with a knife or the teeth. After the blasting cap has been crimped to the fuse, thick grease or tallow (not oil) should be spread over the fuse where it enters the blasting cap. You cannot be sure that everything will work satisfactorily, unless you use a high grade of waterproof fuse and a DU PONT No. 6 (red label), or stronger, blasting cap. Blasting caps are weakened by dampness, so it is always better to use a grade stronger than one which would only just explode the dynamite.

Always cut the end of the fuse squarely across, and always put the fresh cut end in the blasting cap.

The iron or steel bars, for punching the holes in the ground should be a little larger in diameter than the dynamite cartridges, sharpened to a blunt point on one end and at least two feet longer than the depth of the ditch. For making holes in dry ground, where the bar is driven down with sledges or mauls, it is a good plan to have it only long enough to extend about four inches above the surface when the hole is deep enough. This short bar should have a ring or collar welded on close to the top, so that it can be easily loosened with a crowbar if it becomes fast. The bars should have a blunt point and not a long, tapering one, because the dynamite cartridge cannot get to the bottom of a hole made by the latter.

BAR FOR PUNCHING THE HOLES

The labor of punching the holes may be reduced by spacing them farther apart, but when this is done it is necessary to increase each charge of dynamite. If under the conditions described above the holes were spaced three feet instead of two feet apart, the charge of dynamite in each one should be increased to one and a half 1¼ x 8-inch cartridges.

Ditches up to sixteen feet wide can be dug with dynamite, but for this width three rows of holes placed alternately as shown below, are necessary.

```
o   o   o   o   o   o   o   o   o
  o   o   o   o   o   o   o   o
o   o   o   o   o   o   o   o   o
```

The rows should be about the same distance apart as the holes are in the row.

When deeper ditches are wanted, the holes must be put down deeper, and larger charges of explosives be used.

One of the Missouri customers of the Du Pont Company recently stated, that a ditch which he had just dug through swampy land with dynamite, for $100.00, would have cost $400.00 or $500.00 if it had been dug in any other way.

When the clay or gumbo is only moderately wet or damp, it is necessary to locate the holes closer together, if a blasting cap is used in only one of them, as the shock from exploding dynamite will not carry as far through the dryer ground, as it will through wet ground or water; and it will not carry as far through light or loose earth, as through that which is close and heavy. The distance between the holes in clay, which is only damp, should not be greater than a foot and a half; and in wet sand, from a foot to a foot and a half.

When the ditch is to be dug through dry ground, it is necessary to have a detonator (blasting cap and fuse or electric fuze), in each hole, and to tamp with damp earth above the explosives to the top of the hole. The best plan is to fire a number of holes at one time by electricity, so that each charge can help the other.

Although it takes a little longer to dig ditches with dynamite in dry ground than it does in wet ground, because each hole must be

SUBSOIL BLASTING BY ELECTRICITY—HOLES TWENTY FEET APART

primed, it is but little, if any, more expensive as the holes can be spaced from two and a half to three feet apart, and a lower grade of dynamite can be used. The proper explosive for blasting dry ground in this way is Red Cross Dynamite of 25% to 40% strength, about two 1¼ x 8-inch cartridges of which should be used in each hole.

In blasting ditches, as in all other blasting, the grade and quantity of explosives to use, and the spacing and depth of bore holes, are governed by the kind of ground in which the work is to be done, whether it is wet or dry, and by the size of the ditch to be dug.

Railroad builders in order to save handling when grading, often load the bore holes in cuts through shale, or even through rock, so that the material blasted will be thrown as far out of the way as possible, or "wasted," as they call it. This is the principle applied when digging ditches, all of the earth being thrown out and away from the ditch, so that little shoveling is necessary.

There is a great deal of difference between the way an explosive acts in close, wet ground, which offers considerable resistance, and the way it acts in open dry ground, offering very little resistance. Consequently it is necessary, in order to get the best work out of explosives, that is, to make them do the most work possible for the money they cost, to use one kind in close, wet ground, and another kind in open, dry ground. It is also cheapest to space the holes and to fire the explosives differently according to the kind of ground.

As ground in which ditches are to be dug is not always either wet clay or dry loam, but often ranges between the two, it may be necessary to modify the two methods described above in the spacing of bore holes and size of charges.

When shale or rock is encountered in digging ditches, holes must be drilled and fired as in any other kind of rock blasting. The depth and spacing of these holes, and the kind and quantity of explosives which will give best results, depends altogether on the nature of the rock.

DRAINING SWAMPS

In comparatively flat, open country underlaid by a clay subsoil, swamps, often several acres in extent, occupy the lower areas. These swamps are natural reservoirs, generally formed by drainage filling up the depressions until the water is high enough to flow out over the lowest barrier. They are not fed by springs but by the natural drainage. To drain these swamps by means of ditches would sometimes require several miles of ditching, which in many cases is impracticable. Yet such swamps often cover the very best kind of crop-bearing ground, and accordingly any practical method by which they can be disposed of, is a matter of great interest to many farmers. That many of them can be completely drained without ditching, and at very little expense, has been clearly demonstrated. This is done by boring holes down through the impervious ground in the bottom of the swamps, to the sand, gravel or other open material below, and exploding comparatively heavy charges of Red Cross Semi-Gelatin No. 2 (40%) in the holes. The number of holes required depends on the size of the swamps and the thickness of the hardpan. Sometimes a comparatively large swamp can be permanently dried up by means of three or four well-blasted holes in the lowest part of the bottom. Again a row of holes, spaced thirty or forty feet apart, across the lowest part of the swamp may be necessary. For lasting results, the holes must sometimes be drilled as deep as thirty feet, although ten-foot holes are often sufficient. A two-inch wood or dirt auger is generally used. This auger is welded to a piece of gas-pipe six or eight feet long. The top of this is threaded and a tee attached so that a wooden or smaller pipe-handle can be used to turn the auger. When the hole is bored down until the handle of the auger approaches the surface of the water, or as low as it can be conveniently operated from the raft or boat, the tee is unscrewed, another section of pipe attached, and the tee is screwed on to the top of this section. This process is repeated until the hole has reached the required depth. Then the dynamite is loaded. To do this properly a section of two-inch gas-pipe should be pushed into the hole for six or eight inches. This pipe should be long enough to extend above the surface of the water. Through this the 1¼ x 8-inch cartridges of Red Cross Semi-Gelatin should be pushed, one or two at a time, with a wooden stick, to the bottom of the hole, where they will remain in position when the stick is withdrawn if given a good firm push. The last cartridge should be primed with a Victor No. 6 (red label) Waterproof Electric Fuze, for good results cannot be expected if fuse and blasting caps are used in this work. The electric fuze must have wires long enough to reach well above the surface of the water. No tamping is required, as the water above the explosive serves the purpose. After the primer cartridge is in position, draw out the gas-pipe carefully, so as not to pull the electric fuze out of the cartridge. The electric fuze wires should then be connected to well-insulated leading wire (all connections being carefully waterproofed), and the leading wire should be carefully paid out, while the raft or boat is moved fifty or sixty yards away. The other ends of the leading wire should then be attached to the blasting machine, the operation of which will explode the charge in the bottom of the hole. The approximate charge for a hole ten feet deep is seven cartridges, for a hole fifteen feet deep, twelve cartridges, and for a hole thirty feet deep, twenty-five cartridges, 1¼ x 8-inches, of Red Cross Semi-Gelatin No. 2 (40%).

A few years ago a Kansas farmer, who had owned a farm for twelve years with a forty-acre swamp on it, blasted a row of holes spaced about thirty-five feet apart, across the lowest part where the water was three feet deep. The swamp dried up immediately and the first year sixteen hundred bushels of oats were raised on the ground it had occupied. Since then four crops of alfalfa per year have been raised on this ground.

GRADING A ROAD—THE BLAST

ROAD BUILDING

No farm can be properly run without good, hard roads, and the only way to have good roads about the farm is to lay them out correctly in the first place, and then grade and ditch them properly. This grading and ditching always takes more or less digging, but by using dynamite to loosen up the hard ground or shale, and to blast out the rock, roads can be built quickly and at comparatively little expense.

To blast cuts not more than five feet deep through hard earth or shale, drive a bar down to within six inches or a foot of "grade," and in the hole thus made explode one or two cartridges of Red Cross S. P. I. (30%) Dynamite. Be sure to first tamp the charge properly. Holes should be spaced five to eight feet apart. In this way the material to be removed is not only broken up so that it can be shoveled very easily, but a good portion of it is spread over the surrounding land and does not have to be handled.

Roads can be ditched with but little shoveling, by exploding about half a cartridge of Red Cross S. P. I (30%) Dynamite in holes along the sides a foot deep and two to three feet apart.

If it is necessary to cut through rock, the holes should be drilled closer together and charged heavier. When crushing stone for country roads and turnpikes, it will be found much cheaper to locate the crusher at a ledge, and blast out the rock, than to keep teams traveling over the whole countryside, gathering up boulders and hauling them to the crusher.

DIGGING CELLARS AND FOUNDATION TRENCHES

No farmer wants to put an expensive barn or house on a poor foundation, and it would be hard to find a greater benefit to a farm than a good cellar. The proper location for a building is on a knoll, and the rock often comes nearer to the surface on the knolls than it does in the hollows. This makes both good foundations and good cellars possible with the help of dynamite. With sharp drills, a couple of sledges, Red Cross Extra (40%) Dynamite, fuse and No. 6 (red label) blasting caps, a cellar can be excavated, and the rock squared up for foundation or walls, in a short time and at little expense. In fact, it takes little, if any, more blasting to put a house or barn on rock, if same is to be found, than it does to quarry elsewhere the necessary stone for cellar and foundation walls. The rock blasted out can also be used for the walls and piers.

If it is not possible to locate the house and barns on a rock, then the foundation trenches and cellar in the clay, gravel or shale, can be dug much more quickly and easily, if the material is loosened up with an occasional charge of Red Cross S. P. I (30%) Dynamite.

The stone for cement construction of all kinds can be easily blasted out in the proper size for the crusher with Red Cross Extra (40%) Dynamite.

WELL SINKING

Wells are frequently sunk through rock or ground which cannot be dug to advantage without the aid of explosives. In well sinking, when rock is reached and the earth or sand above is properly shored, a circle of four or five drill holes should be started about half-way between the center and the sides of the well and pointed at such an angle, that they will come close together near the center when they are three or four feet deep. These holes should be loaded about half full of 40 per cent. Gelatin Dynamite, with damp clay tamping packed firmly above this to the top of the hole, and then fired simultaneously from the surface by electricity. The result of this shot will be to blow out an inverted cone or funnel in the center, and the bottom can then be squared up with another circle of holes drilled vertically as close to the walls as possible. If the well is large it may be necessary to drill a circle of holes between the inner and outer circle. The above process should be repeated until the well has passed through the rock or has been sunk to the necessary depth. Nothing weaker than No. 6 (red label) Electric Fuzes should be used to explode the Gelatin Dynamite.

DIGGING HOLES FOR POLES AND POSTS

Much time and labor can be saved by using dynamite as an assistant in digging holes for poles and posts. Only enough explosive should be used to make the digging easy; larger charges loosen up the ground to such an extent, that it is difficult to make the poles or posts as firm as they should be. This applies particularly to large poles.

The best way to do this blasting is to drive a crowbar down into the ground within six inches of the desired depth of the hole. In the bottom of the hole made by the crowbar, explode from one-quarter of a cartridge to one cartridge of Red Cross Extra (40%) Dynamite. **Do not tamp,** as this would cause the explosive to loosen the ground too much.

THE SAME FIELD BEFORE AND AFTER CLEARING

THE THIRD STAGE

PREPARING THE LAND FOR CROPS

PLOWING WITH DYNAMITE

DYNAMITE IN THE ORCHARD

SUBSOIL BLASTING—HOLES CHARGED AND FUSE READY TO LIGHT

PREPARING THE LAND FOR CROPS

PLOWING WITH DYNAMITE

PLOWS have been used for many centuries to break up the surface of the earth, so that seeds could be easily planted and grow better; but from the forked stick of the savage to the great gang plow of the modern farmer, plows have rarely broken up the ground satisfactorily to a depth greater than eight or ten inches.

In rich, open soil that is properly fertilized, plowing will break up the ground sufficiently for many years, but eventually a hard stratum, sometimes called "plow hardpan" or "plow sole," forms just below the depth to which the plow reaches. This "plow sole" must be broken up, or the land will begin to fail. Even when "plow sole" does not exist, land that has been bearing crops for many years is wonderfully improved, if the lower soils are turned up and aerated in every six or eight years.

Land that has only a thin surface soil underlaid by an impervious or partly impervious subsoil, sometimes known as "hardpan," will not produce crops at all, or else only produce them in a very limited degree, or for a short time only, unless the subsoil be thoroughly broken up. Where the fertile surface soil is fairly deep, attempts have been made to farm this land by means of drainage and irrigation, the surplus water of the wet season being drained off, and the lack of water in the dry season being overcome by irrigation. This system is expensive, and at best gives only temporary relief, for in the end it tends to increase the thickness of the hardpan, and in some localities may be responsible for alkali soil. The heavy rains of the wet season often cannot penetrate this subsoil, so the water rushes to the low ground, carrying much of the valuable surface soil with it. There, still unable to sink through the hardpan, it prevents plowing until the water evaporates, or causes the seed to rot if it is already in the ground.

SUBSOIL BLASTING—LIGHTING THE FUSE

In some parts of the country, ridges or "dikes" are run at regular intervals over all of the higher ground, to keep the thin surface soil from being washed away. Considerable time and labor is required to make these dikes, as well as to keep them in shape, and they take up ground that should be bearing crops. On land of this kind, drainage, irrigation, dikes, late plowing and rotting seed, could all be done away with if the subsoil or hardpan were properly broken up.

There is only one practical way to upturn and aerate these lower soils, or to shatter hardpan, and that is by blasting.

Subsoil and hardpan can be plowed just as effectively with explosives as surface soil can be with a plow, and just as cheaply, too, for it is only necessary to do this subsoil plowing once in a number of years.

Blasting subsoil has been practiced for some years by a few progressive farmers in different parts of the country. These men have had wonderful results in the way of increased crops on land already under cultivation, and in transforming into excellent growing land that which would have been worthless without dynamite.

It is only lately, however, that the benefits from plowing with dynamite have become generally recognized. The Department of Agriculture, the various agricultural colleges, and men prominently identified with farming in different parts of the country, are all now making a careful study of this question in order to determine the strength, quantity and kind of explosives, the most effective way of using them, and the depth and spacing of the holes, for best results. Farmers everywhere are experimenting on their own responsibility. Many are already claiming that subsoil plowing with explosives will be a common custom in a few years, and will represent millions of dollars in increased crops from lands already under cultivation; and more millions from land now producing nothing but weeds and considered entirely worthless. Along with this increase in land production, will be a reduction in expensive draining and irrigating; because when hard subsoils are properly shattered with explosives, the surplus water falling in rainy seasons sinks down to the lower soils, and does not need to be drained off. Neither is it necessary in dry seasons to provide water

SUBSOIL BLASTING
THE RESULT--CORN RAISED ON THE BLASTED GROUND

by means of irrigating to keep the crops from drying up, because the water which has settled deep in the ground during the heavy rains, is drawn up in the dry season by the plant roots as they need it.

In some places the subsoil is so close to the surface, and so hard, that the plant roots grow down to it but are unable to penetrate it. Ground of this nature will not produce crops amounting to anything until the subsoil has been blasted, no matter how much it is drained and irrigated. After being blasted, however, it often becomes wonderfully productive.

Different methods of loading and spacing the holes are practiced in different places, the usual distance apart in the South being four to eight feet, with a charge of about one-quarter of a 1¼ x 8-inch cartridge of Red Cross S. P. 1 (30%) or Red Cross Extra (40%) Dynamite in each hole; in Kansas and other Western States, fifteen to thirty feet apart, with a charge of from one to one and a half 1¼ x 8-inch cartridges of Judson Powder FF (15%), or Red Cross S. P. 3 (25%) Dynamite in each hole. The proper place to locate the charge so that it will shatter a maximum area of subsoil, is two to three feet below the surface if the hardpan is thin, and the holes are placed four to eight feet apart. In thick hardpan with the holes fifteen to thirty feet apart, the charge should be located about a foot above the bottom of the hardpan.

In California, hardpan has been blasted very effectively and cheaply with Judson Powder R. R. P. (5%). A slow-acting dynamite does better work than a quick one, because its effect is more spreading. The difference in the results of the use of slow explosives and of quick ones on hardpan is illustrated on pages 74, 76 and 78.

The exact size of the charge, the proper spacing of the holes, and their depth, depend largely on the kind of subsoil to be broken, its thickness and depth below the surface.

When the bore holes are fifteen feet or more apart, it is best to explode a number of charges simultaneously by means of electricity, but when the bore holes are closer, it answers the purpose about as well to explode the charges with fuse and blasting caps.

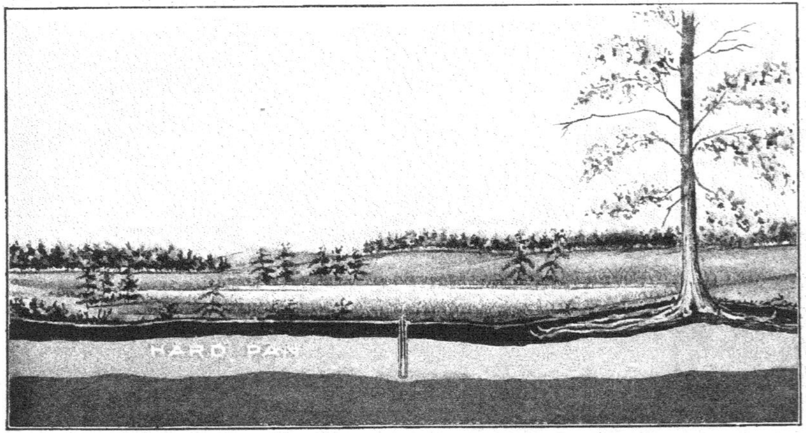

SECTION THROUGH HARDPAN, SHOWING PROPER DEPTH FOR BORE HOLE

Land owners and farmers in Kansas and in the South, fruit growers in the Western States, and others, have profited enormously in the last few years by plowing subsoil with dynamite, and these men are its greatest advocates.

In Kansas it has been the custom to use only fifty or sixty pounds of Red Cross Dynamite for blasting an acre of ground, and to keep the cost of explosives and labor below ten dollars per acre; but the Southern farmers insist that it is an excellent investment to spend three or four times as much per acre.

As proof of the benefit of plowing with dynamite, F. G. Moughon, of Walton County, Georgia, refers to the phenomenal yield of watermelons, averaging fifty to sixty pounds each, which he has raised for years on land blasted with charges of about one-third of a 1¼ x 8-inch cartridge of dynamite in bore holes, two and a half to three feet deep, spaced eight and ten feet apart. J. H. Caldwell, of Spartanburg, South Carolina, advocates holes two feet deep spaced four feet apart. He explodes in each of these about one-quarter of a 1¼ x 8-inch cartridge of Red Cross S. P. 1 (30%) Dynamite. These holes are not thoroughly tamped, but when they are loaded enough loose dirt is pushed into each one to close up the opening and hold the fuse upright. Although this lack of tamping undoubtedly results in wasting some of the force of the dynamite, Mr. Caldwell believes that it is cheaper in the end, because of the saving of time in loading the holes. When the ground was not blasted the corn had to be planted eighteen inches apart in the row, with rows four feet apart. The blasted ground was able to nourish stalks six inches apart, with the rows spaced as before. The corn on the blasted ground grew almost one-half taller than that on the unblasted ground and produced a crop 33% greater. The corn was weighed by a committee of the Chamber of Commerce, of Spartanburg, South Carolina, who also report that the blasted ground produced fuller ears and firmer grains.

It is impossible, in the limited space of this handbook, to give details concerning the results of the subsoil blasting done by many of those who have profited greatly by it, but J. T. Garrett, of Laurens,

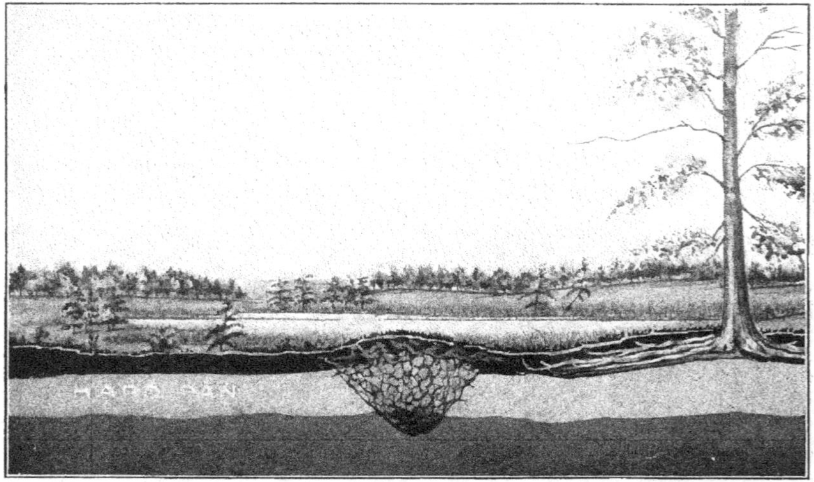

SECTION THROUGH HARDPAN, SHOWING EFFECT OF A QUICK-ACTING EXPLOSIVE

North Carolina, who wonderfully improved his corn and watermelon yield, and M. T. Williams, of Medicine Lodge, Kansas, who increased the value of his land **almost tenfold** for alfalfa, must be mentioned.

In many parts of California, and some of the other Western States, true hardpan exists. The only way in which land underlaid by this hardpan can be made worth anything, is to break up the hardpan with explosives. Then fruit trees will grow and bear just as well as anywhere else.

Tools Used when Blasting Subsoil

In the South, the holes in which the dynamite is exploded are made by driving into the ground with sledges a steel bar, two and a half to three feet long, and a little larger in diameter than the dynamite cartridges. This bar has either a flange or a lug near the top, so that after it has been driven down it can be loosened with a crowbar. The point should be sharp but blunt, for a long, tapered point would make a hole so small at the bottom that the dynamite cartridge would lodge part way down. An illustration of this bar will be found on page 56.

In Kansas the holes are usually bored with a two-inch dirt or wood auger having a long shank, as the bottom of the hardpan there is often several feet below the surface. A crowbar to drive stones out of the course of the auger is also useful.

If you have not tried blasting subsoil, or "plowing with dynamite," we suggest that you lay off an acre, or else a plot 100 feet by 200 feet in one of your fields, and blast it as described above. Plant the whole field and note how much better and heavier the crop is on the blasted ground. An experiment of this kind will cost but little, and should be the means of greatly increasing your income in the future.

SECTION THROUGH HARDPAN, SHOWING EFFECT OF A SLOW-ACTING EXPLOSIVE

DYNAMITE IN THE ORCHARD

That dynamite is an exceedingly valuable aid in the successful growing of trees can no longer be doubted. Those who have tried it, are firmly convinced that no method of excavating the hole for the roots of the young tree, whatever may be the soil conditions, is so economical, quick or productive of as desirable after-results, as blasting with dynamite.

Dynamite should be used when planting trees, because, one cartridge of Red Cross S. P. I. (30%) Dynamite will excavate instantly as large a hole in which to plant a tree as a man will dig in an hour; and because the explosion of dynamite loosens up the soil for many yards around, giving the tree roots a better opportunity to spread out than when the hole is dug by hand. The explosion also destroys all insects and grubs in the ground nearby. Every year or so during the life of the trees, small charges of Red Cross S.P.I.(30%) Dynamite should be exploded midway between them, and some four or five feet below the surface of the ground. This tends to keep the soil open so that it will hold moisture and gives the tree roots every chance to spread, besides keeping the ground free from grubs.

When older trees begin to fail it is sometimes of much benefit to detonate a charge of explosives directly under them. To do this a hole should be started seven or eight feet away from the tree and driven downward and toward the tree on a dip of about 45°. When the hole is in twelve feet, the bottom will be from eight to nine feet directly under the trunk of the tree. This hole should be "sprung," or "chambered," with a 1¼ x 8-inch cartridge of Red Cross Extra (40%) Dynamite, and then loaded with from five to ten pounds of Judson Powder R. R. P., with tamping above the charge to the mouth of the hole. The explosion of this charge breaks up the hard soil below the roots of the tree, so that it can hold a greater quantity of moisture which the tree roots will take up as they require it. The beneficial results from this blasting will not appear in the succeeding crop, but will be manifest in the second and third crops afterward.

CROP OF WATERMELONS GROWN ON LAND PLOWED WITH DYNAMITE

When breaking up hardpan between the trees in the California orange groves, it is the custom to bore down with an auger just through the hardpan—usually about four feet below the surface. This hole is "sprung" or "chambered" with one-half of a 1¼ x 8-inch cartridge of Hercules Stumping Powder (10%), and then, after being loaded with about one and a half pounds of Judson Powder R. R. P., and well tamped, is blasted.

If the soil is inclined to be swampy, heavier charges, detonated deeper in the ground, will break up the lower impervious stratum, and permit the surplus water to sink into the earth, where it will be conserved, and afterwards given up to the roots of the trees as they require it.

An example of the great benefit derived from the use of explosives, for excavating the holes in which young trees are to be planted, was recently brought to our attention by a well-known fruit grower, who reported that he planted nine peach trees a few years ago to determine positively whether anything was to be gained by using dynamite. Three of the trees were planted in holes made by drilling a two-inch auger hole three to four feet deep, and detonating a charge of explosives in the bottom; the other six trees were planted in holes of the regulation size dug by hand. Three years later the three trees which had been planted in the blasted holes were strong and healthy, and produced between five and six bushels of very fine peaches; but the other six trees planted on the same ground without blasting, bore practically no peaches at all, both fruit and leaves having shrivelled up and dropped off during the dry season.

A similar experience was that of a Western farmer, who set out an apple orchard more than twenty years ago. After digging a number of holes for the young trees, he decided that life was too short for such slow methods and drove to town where he bought a case of dynamite, some blasting caps and fuse from the hardware dealer. With this he blasted holes for the remaining trees and today, twenty years later, he reports that the trees planted in the blasted holes, are superior in every way to the others, and that they have produced better fruit, and more of it, ever since they began bearing.

DU PONT DEMONSTRATOR SHOWING HOW TO PRIME A DYNAMITE CARTRIDGE FOR HARDPAN BLASTING

One successful Missouri fruit-grower states that he would never think of replanting where a tree had died out, without blasting the hole for the new tree with dynamite. The new trees then are not so likely to be attacked by disease, and are markedly better than those planted in holes which are not blasted.

So many other similar cases have been brought to our attention, that we feel every fruit grower should give the question of using explosives careful and thorough consideration.

A GOVERNOR, A CONGRESSMAN, AND A STATE SENATOR, WITNESSING A DU PONT HARDPAN BLASTING DEMONSTRATION IN KANSAS

MAKING HAY FOR THE CHURCH ON A DYNAMITED FIELD WHERE FORMERLY LITTLE OR NOTHING GREW

A BAPTIST CHURCH PICNIC IN KANSAS

(Reproduced by kind permission of "Farmers Mail and Breeze" Topeka, Kansas.)

A few years ago M. T. Williams bought a quarter section of land near Medicine Lodge in Barber County and, conceiving the same idea that ex-Governor Crawford and others have, used dynamite in dealing with a hard subsoil. The land was overgrown with sunflowers and cockleburs and would have been considered dear at $10 per acre. It was underlaid with a hard subsoil that was almost impervious to water. Mr. Williams' idea was to loosen this subsoil with dynamite. He bored holes in the earth some 3 feet deep and about 40 feet apart, and in each hole placed a part of a stick of dynamite. The explosion of the dynamite loosened the hard subsoil and made a reservoir for the rains which had formerly run off the land nearly as fast as they fell.

THE LAST STAGE

KEEPING UP THE FARM

BREAKING UP ICE GORGES

STARTING LOG JAMS

BREAKING UP LOG ROLLWAYS

BORING HOLES FOR HARDPAN BLASTING IN KANSAS

KEEPING UP THE FARM

AFTER a farm has been carved out of the wilderness, has been properly laid out and drained, with all buildings erected, roads graded and fences up; when wells have been driven, fruit trees planted and subsoil blasted, there is still work to be done year by year, and in much of this work the use of dynamite means a great saving of time and money. Its use for blasting up subsoil in land that is deteriorating, for blasting under and between failing fruit trees, and for cleaning up occasional stumps and boulders overlooked in the first clearing, has already been referred to. It is also necessary for keeping roads and ditches in shape, and is indispensable in those parts of the country where ice may form gorges in the streams in the early Spring. Unless these gorges are promptly broken, great damage may be caused by the water backing up behind them and sometimes overflowing farms and towns, carrying away buildings and bridges.

Dynamite is also of great importance in starting log jams, breaking up "rollways," etc., in the districts where timbering is carried on.

BLASTING ICE

Ice in streams sometimes forms jams or gorges ten to forty feet high. When water backs up behind these, bridges may be carried away and other great damage and loss to the community result. All of this can be prevented at a small expense by the use of DU PONT Dynamite.

To break up floating ice so that a gorge will not be formed, charges of dynamite should be exploded on the surface of the ice, the size of the charge depending on the thickness of the ice. If the floating ice is in large sections, the work of blasting should be conducted on a broad, slow-running part of the stream, where it is possible to get on

A PLACE WHERE A FEW DOLLARS' WORTH OF DYNAMITE USED AT THE RIGHT TIME MIGHT HAVE SAVED THOUSANDS OF DOLLARS to the ice either from the shore or in boats. Successive charges, consisting of a number of cartridges of Red Cross Semi-Gelatin No. 2 (40%) tied together in a bundle, should be laid on the ice and exploded with fuse and blasting cap, until the ice is properly broken. When the streams are narrow the charges of explosives may be thrown on to the ice from the shores or, if the ice is running swiftly, the charges may be dropped on to the cakes from the down-stream side of bridges. The charge to be thrown on the floating ice should be prepared by tying securely together in a bundle, the required number of cartridges of Red Cross Semi-Gelatin No. 2 (40%), the cartridge in the middle of the bundle having been primed with a blasting cap and waterproof fuse. The place where the fuse enters the blasting cap, should be well coated with soap or thick grease to keep water from getting into the blasting cap and damaging it. A block of wood, a stone, or some other object that would prevent its rolling should then be tied to the charge which, after the fuse is lighted, should be thrown or dropped as nearly as possible on to the middle of the ice cake. Particular attention must be given in this operation to the length of the fuse. Fuse generally burns from two to three feet per minute, and when the fuse is lighted with the dynamite still in the hands of the operator, extra time for possible emergencies, should be allowed.

To blast out ice gorges the charge, usually of from five to twenty-five pounds of Red Cross Semi-Gelatin No. 2 (40%) should be pressed into a hollow or crevice at what appears to be the weakest part of the gorge. This charge should be exploded from a distance by electricity, so that the operator can be on shore when the explosion takes place and the gorge moves out. If the first shot does not start the gorge the process should be repeated until it is broken. To open great ice gorges, such as the one that formed in the Niagara River in April, 1909, much larger charges are necessary. In one blast on this work a charge of two thousand five hundred pounds of Red Cross Dynamite was exploded.

Watering places for stock along the banks of streams can be easily kept free from ice in winter by the use of a little dynamite from time to time.

Since ice blasting is always done in cold weather, it is necessary to give particular attention to having the dynamite thoroughly thawed. If this is not done poor results will follow. Red Cross Dynamite is always recommended for this work, because it is not so easily affected by cold weather as other standard brands, and can be thawed more quickly.

"STARTING" LOG JAMS

Red Cross Semi-Gelatin No. 2 (40%) is invaluable for breaking log jams. A charge exploded on a log, above or below water, will cut it in two as readily as can be done with an axe, with the advantage of the operator being at a safe distance when the jam "starts."

BREAKING UP "ROLLWAYS"

Red Cross Semi-Gelatin No. 2 (40%) can be used to great advantage in breaking up "rollways" of logs, large quantities of this powerful explosive being used each season by the lumbermen. A case of this explosive, thawed in the morning, will, if covered by an old coat or blanket and protected from the cold, be ready for use at any time during that day. The huge piles of logs, frozen and bound together with snow and ice, can be instantly loosened with a little dynamite instead of prying them apart slowly and laboriously with cant hooks and levers. At a season when time is truly money, the dynamite used in this way saves many times its cost. The cases of dynamite can be transported easily and safely if properly handled. No logging camp can afford to be without it a single day when engaged in this work.

EXPLOSIVES AND BLASTING SUPPLIES

HOW TO HANDLE, STORE AND USE THEM

BLASTING BY ELECTRICITY

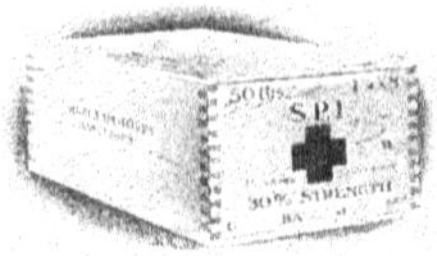

50-lb. Case Red Cross S. P. 1, 1¼x8"

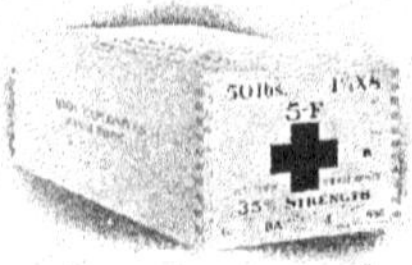

50-lb. Case Red Cross 5 F, 1¼x8"

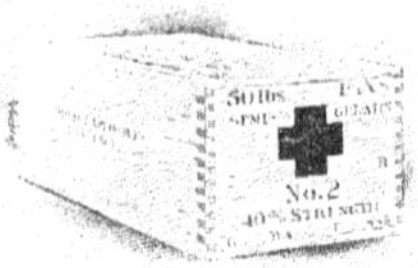

50-lb. Case Red Cross Semi-Gelatin No. 2, 1¼x8"

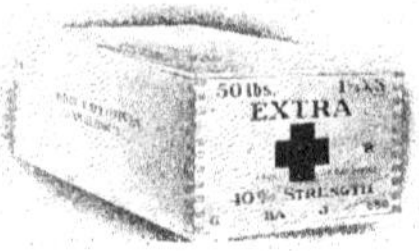

50-lb. Case Red Cross Extra, 1¼x8"

50-lb. Case Hercules L. F. Stumping Powder, 1½x8"

50-lb. Case Hercules No. 1, 1¼x8"

50-lb. Case Judson R. R. P., 4 bags

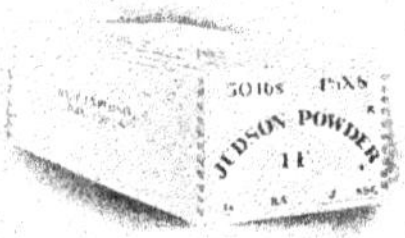

50-lb. Case Judson FF, 1¼x8"

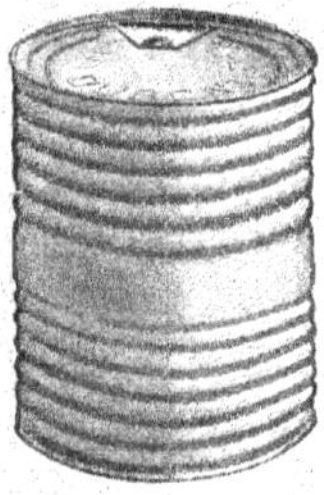

25-lb. Steel Keg FF Blasting Powder

25-lb. Pulp Keg FFF Blasting Powder

PRINCIPLE OF EXPLOSIVES

BLASTING Explosives are divided into two general classes, known as High Explosives and Low Explosives; Red Cross Dynamite and Judson Powder are examples of the former, and Blasting Powder of the latter. They are solids, having bound up in themselves very powerful energy which, when properly directed, can be made to do an enormous amount of valuable work. To get them to do this work economically, they must be closely confined in the midst of the material which is to be broken or thrown out, and then exploded, that is, instantly changed from a small volume of a solid to a very large volume of a gas. Low Explosives (Blasting Powder) will do but little, if any work, if not tightly confined or corked up, when exploded, but High Explosives change from a solid to a gas so quickly, that they will jar and break material on which they are lying when they explode, even if they are confined very little, as in "mudcapping" or "blistering" boulders, or if not confined at all, as when floating ice is blasted.

Low Explosives are exploded by a spark, but a spark will not explode High Explosives, although it may ignite them, and the heat and pressure caused by burning in a confined space may result in an explosion after a time.

High Explosives can only be properly exploded by a very powerful shock. This shock is brought about in their use by exploding a detonator which has been inserted in the charge of explosives. This detonator is either a blasting cap, which is exploded by a spark from the fuse, or an electric fuze (pronounced fu-zee) which is exploded by a fine wire superheated by an electric current.

When fuse and blasting caps are used it is necessary for the blaster to cut the necessary length of fuse from a roll, and fasten the blasting

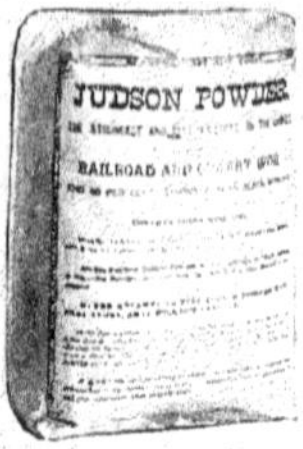

12½-lb. Bag of Judson Powder R. R. P.

1¼x8" Cartridge of Red Cross S. P. 1

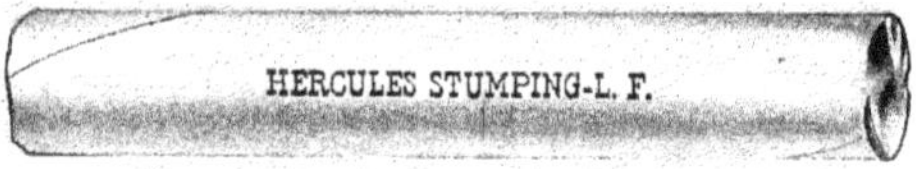

1¼x8" Cartridge of Hercules Stumping L. F. Powder

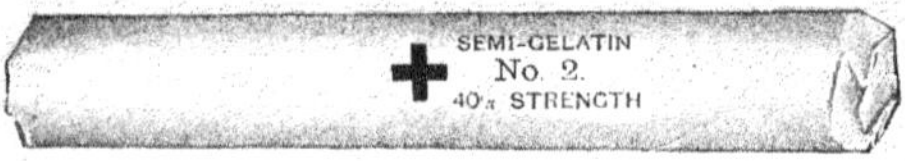

1¼x8" Cartridge of Red Cross Semi-Gelatin No. 2

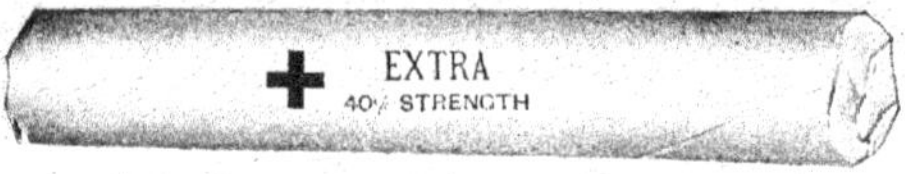

1¼x8" Cartridge of Red Cross Extra

1¼x8" Cartridge of Red Cross S. P. 3

cap to the freshly cut end of the fuse with a cap crimper. The electric fuze is furnished with the two wires sealed into it.

Blasting Powder is exploded by the spark from fuse or an electric squib, no detonator being required.

The Detonator

When "High Explosives" were first manufactured, they were not as safe to handle or use as the properly-made ones now are. Consequently they could be exploded with a small and weak detonator. The standard explosives now manufactured cannot be **completely** exploded, and give best results, if anything weaker than a No. 6 (red label) detonator is used with them. Nevertheless, many dealers will sell to inexperienced consumers No. 3 or No. 4 detonators, simply because the price is a little lower than that of stronger ones. This they do regardless of the fact that the reduction in price is much more than offset by the loss of power due to the incomplete detonation of the explosive with which these weak detonators are used.

Judson Powder R. R. P., is granular, and to give best results should be detonated with dynamite of not less than 40 per cent. strength, the quantity of dynamite necessary depending on the size of the charge of Judson Powder. (See page 37.)

Storage

Explosives should be stored in a dry, well-ventilated place not warmer than 80° or 90° F. They should always be kept under lock and key, so that children, or irresponsible people will not have access to them, and should not be stored in a locality where hunting or other shooting may be done, unless they are kept in a bullet-proof building.

Most High Explosives freeze at a temperature between 45° and 50° F., and when frozen will explode either only imperfectly or not at all. **Frozen dynamite should therefore never be used under any circumstances.**

Red Cross Explosives and Hercules Stumping L. F. Powder, belong to a class of explosives that do not freeze as readily as other

A DIFFICULT PROBLEM

Although the tree had been cut from this stump for many years, the stump was still comparatively solid.

standard explosives. They are particularly adapted to blasting about the farm, owing to the fact that they do not freeze until water freezes and thaw when ice melts. When exposed to extremely low temperatures they freeze more slowly, and if warmed thaw more quickly, than any other dynamite. Therefore, when they are used, the time, trouble and possible danger attached to the thawing of dynamite is reduced to a minimum. To the farmer who, perhaps has only an occasional call to use High Explosives, Red Cross Dynamite and Hercules Stumping L. F. Powder are invaluable. On large operations where a considerable amount of dynamite is used, the great saving in time, trouble and expense, as well as the reduction of danger in thawing to a minimum, are points that cannot be overlooked.

Packages and Transportation

Du Pont High Explosives (except Judson Powder, which is usually put up in paper bags) are put up in cartridges, generally 1¼ x 8-inch, and then packed with sawdust in wooden cases. They can, when thus packed, be shipped on freight trains or carried in wagons with but little danger of explosion. There have been instances where they have been dropped, and the cases smashed, without an explosion resulting, showing that these High Explosives stand very rough treatment; still **it should never be forgotten** that they are **explosives,** and we would particularly direct the reader's attention to page 113 of this catalogue, where stress has been laid on "What Not to Do." If these instructions are observed, the user need have no fear of an accidental explosion.

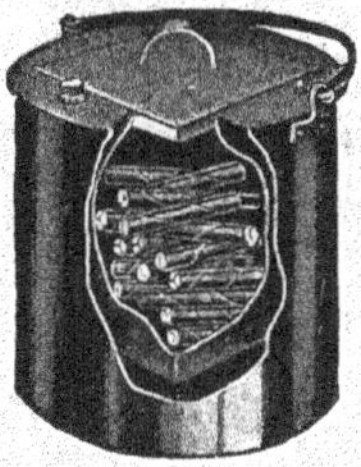

Catasauqua Thawing Kettle

Thawing

There are various ways of thawing High Explosives but the only safe methods are those which **thaw slowly and gradually.**

Burying dynamite in water-tight boxes in manure is a safe and fairly effective way to thaw it, provided the manure is fresh.

EASILY SOLVED WITH DU PONT DYNAMITE
The stump blown entirely out of the ground and split into several pieces

The safest and best way is in a "thawing kettle." A thawing kettle is a double galvanized iron bucket having an outside space for hot water, and an inner water-tight receptacle for the dynamite. The Catasauqua Thawing Kettles are made in one piece, while the Bradford Thawing Kettle is in effect two pails, the outer one for hot water and the inner one for the dynamite. (See illustrations.)

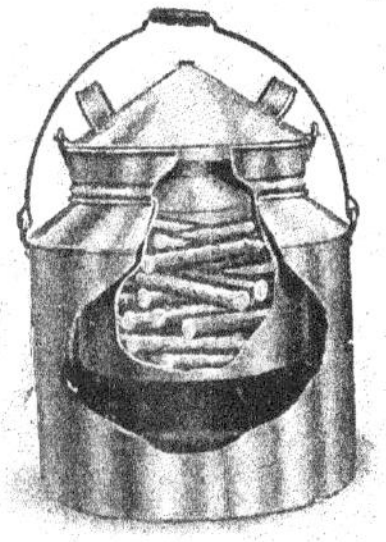
Bradford Thawing Kettle

Never thaw dynamite by putting it near a fire, nor in the oven, nor against a stove or steam pipe. Do not try to thaw by exposing dynamite to steam nor by soaking it in hot water.

It is not at all necessary that the cartridges should feel warm. All that is necessary is to have them soft **all the way through.**

Every user of dynamite who has occasion to thaw it, should make it a rule never to heat water in the thawing kettle, but in some other receptacle, and then after removing the dynamite, pour the hot water into the water compartment. This water should not be so hot that it would burn the hand. The Miners' Thawing Kettle will protect dynamite from the cold better, and keep it in good condition longer, than either of the others, but its capacity is small. This kettle would soon be damaged if an attempt were made to heat the water in it. Water may be heated in the outer pail of the Bradford Thawing Kettle provided the inner pail has been removed, but it is never safe to put a one-piece thawing kettle over a fire, even after the dynamite has been removed, because enough nitroglycerin may have leaked out from dynamite previously thawed, to cause an explosion if the thawing kettle is put over a fire.

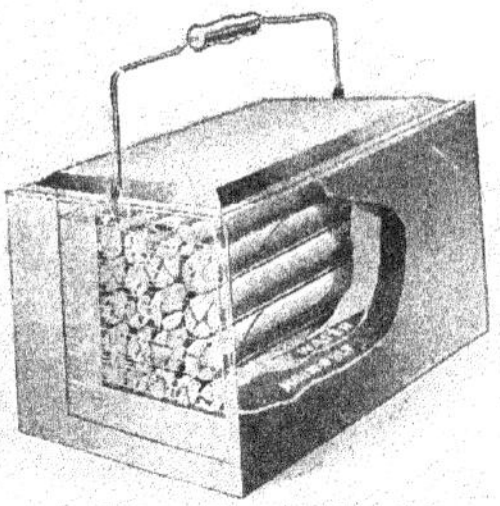
Miners' Thawing Kettle

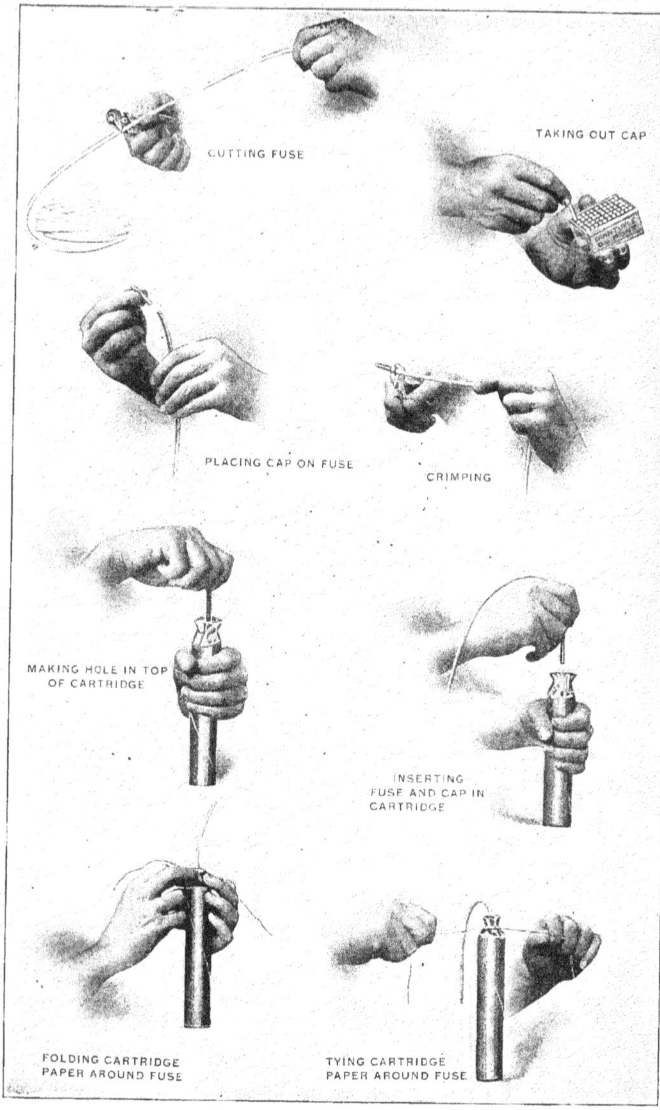

ONE METHOD OF MAKING A PRIMER WITH BLASTING CAP AND FUSE

A simple and effective thawing arrangement may be made by suspending a small pail or bucket containing the dynamite, in a larger bucket partly filled with warm water. Care must be taken to prevent any of the water from getting into the inner pail, which should have a tight-fitting lid. The whole should be covered with a piece of carpet, or an old coat, until the dynamite has thawed.

Priming

Placing the detonator in a cartridge or charge of high explosives is called priming it, and the cartridge or part of a cartridge with the detonator in it is called the "primer" or "primer cartridge."

The first step in the preparation of the primer is to cut the necessary length of fuse from the roll, cutting it squarely across and not diagonally. After carefully inserting the fresh cut end as far as it will go into the blasting cap, fasten the latter securely to the fuse with a cap crimper. When crimping the blasting cap to the fuse, the crimp should be made near the end which the fuse enters so as not to disturb in any way the explosive which the blasting cap contains. An attempt to crimp the blasting cap near the other end would be likely to cause it to explode. The crimp should be made secure enough to prevent the fuse from pulling out of the blasting cap, during the charging and tamping of the bore hole, and, what is quite as important, particularly in wet work, the crimp should be tight enough to keep water out of the blasting cap. A coating of soap, tallow or thick grease, spread over the fuse where it enters the blasting cap will help greatly to keep the water out. This grease should not be applied until after the blasting cap has been crimped to the fuse. Oil should not be used for this purpose as it may soak into the fuse and damage it.

Be sure to cut the fuse long enough to allow it to extend several inches from the mouth of the bore hole when the primer is in place, and also long enough for the blaster to reach a place of safety before the charge explodes. Fuse burns from two to three feet per minute.

To prime a dynamite, or other high explosive cartridge, with blasting cap and fuse, make a hole in the end of the cartridge after

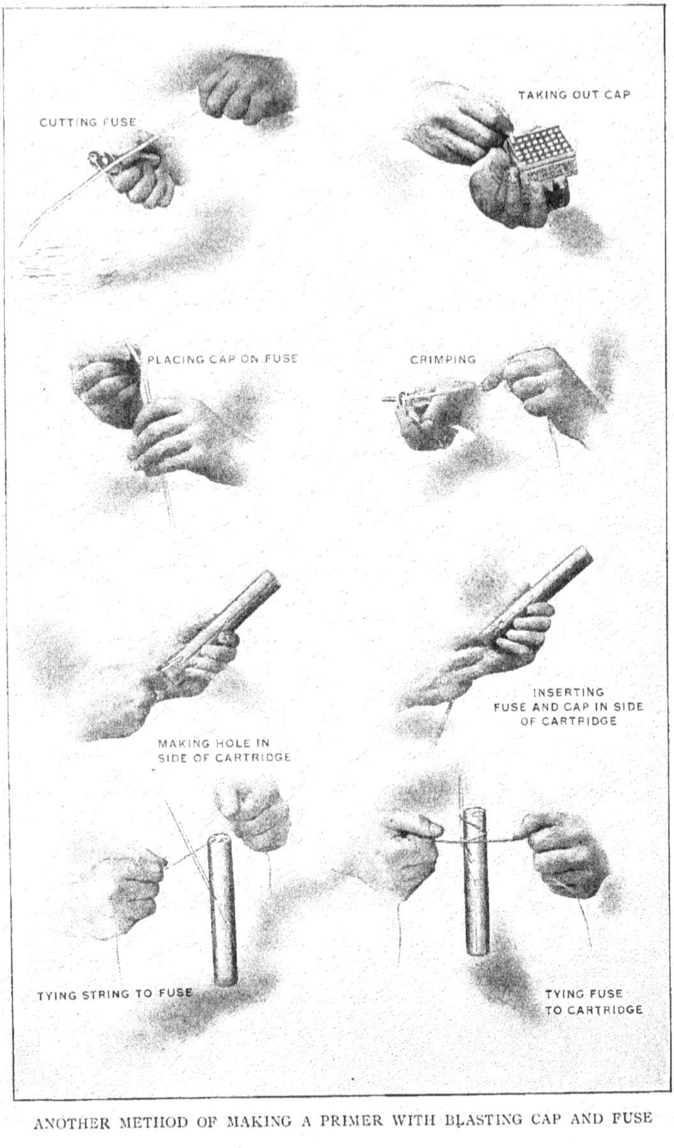

ANOTHER METHOD OF MAKING A PRIMER WITH BLASTING CAP AND FUSE unfolding the paper shell, or in the side of the cartridge near one end, with a small pointed stick, about the diameter of a lead pencil. This hole should not be much larger in diameter than the blasting cap, for an air space around it always detracts from the force with which a blasting cap shoots into the explosive. The blasting cap should not be so deep in the cartridge, that the fuse will come in contact with the explosive for any appreciable distance, as side spitting of the fuse usually ignites the explosive.

Best results will be had if the blasting cap is pointed **straight down** into the primer cartridge.

When the blasting cap has been put in the end of the cartridge, the paper must be folded carefully about the fuse, and tied securely with a piece of string. When the blasting cap is inserted in the side of the cartridge near the end, the fuse is held in position by tying it with a double loop of string around the cartridge. Both of these methods of priming are clearly shown in the illustrations on pages 100 and 102.

Charging

Having primed the cartridge in the manner described, insert it in the bore hole and push it carefully home. It is sometimes well in **dry ground,** to slit the paper shells lengthwise before putting the cartridges into the bore hole. Push them firmly into place so that they will fill up the diameter of the hole, for crevices or air spaces greatly lessen the power of an explosive. If more than one cartridge is used in a charge, it is only necessary to prime the top or outside one, but in the bore hole each cartridge must touch the one previously loaded. If any space between the cartridges occurs through falling dirt or stones, or through the sticking of a cartridge in the bore hole, a partial misfire may result.

Tamping

After the charge is pressed home, as directed, put in two or three inches of fine dirt or sand, and tamp (pack) **lightly.** Then fill up two or three inches more of the hole with tamping material, packing it in a little more firmly. After five or six inches of tamping covers the charge, the tamping may be pressed firmly into place without danger of premature explosion. The tamping material should be packed as firmly on top of the charge as can be done without moving the electric fuze or blasting cap and fuse in the primer, but it is not safe to tamp by a blow any stronger than can be given by hand. Fill the bore hole up with tamping until even with the surface. The firmer and harder the tamping can be made (without overlooking the above precautions) the better will be the results. If the bore hole is not properly tamped, the charge is likely to "blow out," or at any rate some of its force will be wasted.

Be sure the tamping is done with a wooden stick. **Never use a metal bar or anything having metal parts.**

NOTES

When you buy explosives get nothing but the best. Remember that dynamite is nothing but **Concentrated Power,** and when you buy a case of dynamite you are getting a hundred or so **Little Giants** who are going to do a great deal of work with one blow. You cannot see **strength,** you can only perceive its effects. And therefore, when you buy, your only safe way is to get something others have tried and found good. Nearly a thousand different kinds of explosives have been put on the market, but those that have survived and are known to be good can be counted on one's fingers. All others have fooled the people for awhile and then disappeared from the market, though in the first place each was sold as being "just as good" as DU PONT Explosives.

Always be careful when using dynamite, not to taste it nor get it on the hands, for the smallest quantity will nearly always cause a violent headache. It is an excellent plan to wear a pair of old gloves, which should be destroyed before they become saturated with the nitroglycerin in the dynamite. With a little care, a great deal of dynamite can be handled before it is necessary to destroy the gloves.

BLASTING SUPPLIES

When detonating high explosives with fuse and blasting caps, the only supplies necessary besides the fuse and blasting caps, and the tools for making the hole and for tamping, are the cap crimper and thawing kettle. The latter has been described under the paragraph on "Thawing," page 99.

When the blasting is done by electricity, a blasting machine, coil of leading wire and electric fuzes, take the place of fuse and blasting caps. A spool of connecting wire and roll of insulating tape should also be provided. A leading wire reel is not a necessity, but saves time and keeps the leading wire in good condition.

Fuse

Fuse is made in several different grades, put up in double coils consisting of two single coils of fifty feet each, and packed in cases containing from five hundred to six thousand feet. In wet work, "double tape" or "triple tape" fuse should be used, while "single tape" or "cotton" will answer in dry work. Fuse should be stored in a cool, dry place to keep it from hardening and "breaking" when it is unrolled, or from getting soft and defective. In cold weather fuse becomes stiff and must be unrolled very carefully to keep it from breaking. Most fuse burns from two to three feet per minute.

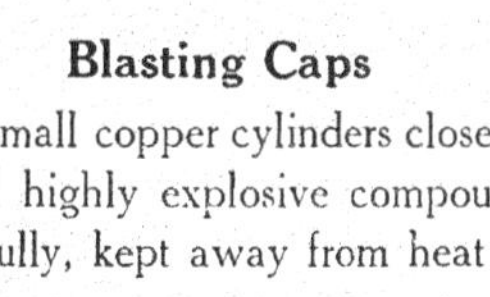

COIL OF FUSE

Blasting Caps

Blasting caps are small copper cylinders closed at one end, which contain a sensitive and highly explosive compound. They must always be handled carefully, kept away from heat or sparks, and must not be subjected to heavy knocks or jars. They should never be carried loose in the pocket, nor permitted to lie about where children may find them. They are very dangerous in the hands of a child or irresponsible person. They should never be stored or carried with dynamite, because they are more easily exploded by accident than is dynamite, and if they explode in close proximity to dynamite, they will probably cause it to explode also, and do great damage. If blasting caps are to do good work, they must be stored in a perfectly dry place, and must not be permitted to lie in a damp place for even a short time before using. The charge which they contain is very quickly damaged by moisture, and although it may explode when damp, it detonates with so little force that it may not explode the charge of dynamite, or may only partly explode it.

Blasting caps are put up in tin boxes containing one hundred each. These boxes are then packed in wooden cases containing from five hundred to five thousand blasting caps. They are manufactured and sold in six sizes—No. 3, No. 4, No. 5, No. 6, No. 7 and No. 8, but nothing weaker than the No. 6 (red label) can be depended on to develop the full strength of the explosive with which they are used. No. 6 Blasting Caps vary a little in length, but are not shorter than 1 3/8 inches, while the No. 5 (blue label) size is usually 1 1/4 inches long, and the No. 3 (Silver Medal) and No. 4 (Gold Medal) sizes, 1 inch long. It is a good rule never to buy a blasting cap shorter than 1 3/8 inches or too small in diameter for tape fuse to fit snugly into it.

DU PONT NO. 6 BLASTING CAPS

Cap Crimpers

The two styles of cap crimpers which are most reliable are known as the DU PONT Cap Crimper and the California Cap Crimper.

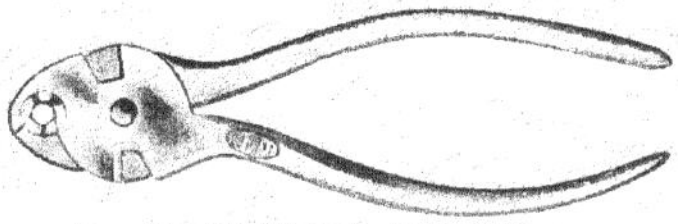

DU PONT CAP CRIMPER

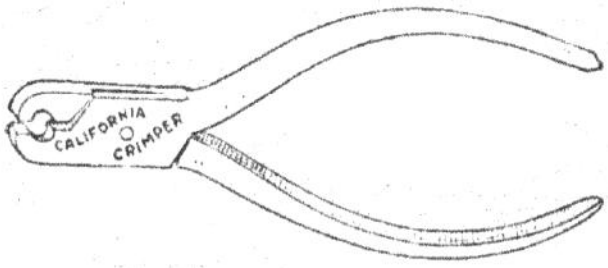

CALIFORNIA CAP CRIMPER

A cap crimper is inexpensive, and absolutely necessary if the blasting cap is to be properly fastened to the safety fuse. Fastening blasting caps to fuse with a knife or with the teeth, is exceedingly dangerous and inefficient.

BLASTING BY ELECTRICITY

When a good deal of blasting is to be done, it is usual to set off the blast by means of a blasting machine and electric fuzes. In this way any number of charges from one up to forty or more can be fired simultaneously. This is found very convenient when it is necessary to get rid of a large stump or an extra big boulder, as frequently several charges fired at the same time will do more work than the same charges fired successively, or than one charge as big as all of the smaller ones combined.

A very little thought will show that more work will be done when firing a number of blasts **simultaneously,** than when firing them singly, while a little experience will teach that much is to be gained by firing even single shots electrically. Another advantage gained in firing by electricity is, that in case of a misfire (which can rarely happen by this method), it is unnecessary to wait a considerable time before returning to the work, as must be done when using safety fuse.

Directions

Use electric fuzes, leading and connecting wire, and blasting machine instead of fuse and blasting caps. The electric fuze wires must be long enough to extend several inches out of the bore hole, and care must be taken when tamping not to break these wires or damage the insulation on them. If more than one charge is to be fired at one time, separate the two electric fuze wires extending from each hole and connect one of these wires to one extending from the hole on one side, and the other one to one of the wires extending from the hole on the other side, and so on, until all are connected together in a string with one free wire extending from the first hole, and another extending from the last hole. If the electric fuze wires are not long enough to reach each other, use a piece of connecting wire to join them.

Be sure that all wire ends are scraped clean and bright before they are connected together.

The charges having all been connected as directed above, the free wire of the first hole should be joined to one of the "leading" wires, and the free wire of the last hole to the other leading wire. The leading wires should be long enough to reach a point at a safe distance from the blast—say two hundred and fifty feet, at least.

All being ready, and not until everybody is at a safe distance, connect the leading wires to the binding posts on the top of the blasting machine, through each of which a hole is bored for the purpose, and bring the wing nuts down firmly upon the wires.

Now, to fire: If using a push-down blasting machine, take hold of the handle, lift the rock bar (square rod toothed upon one side) to its full height and push it down with all force, bringing it to the bottom of the box with a **solid thud,** and the blast will be made.

The Reliable Blasting Machine

This is a push-down blasting machine made in three sizes: No. 2, which will fire up to 20 electric fuzes at one time; No. 3, which will fire up to 30 electric fuzes at one time, and No. 4, which will fire up to 50 electric fuzes at one time. The No. 3 size is made with three posts when so ordered, without extra cost, and the No. 4 size is always made with three posts.

Blasting machines with three posts can be used with either two or three leading wires. When used with two leading wires, one wire is connected to the middle post and the other to either one of the outside posts. When used with three leading wires, those from the two outside posts are connected to the electric fuzes on the ends of the circuit, and that from the middle post is connected to the electric fuze wire, or the connecting wire, somewhere near the middle of the circuit.

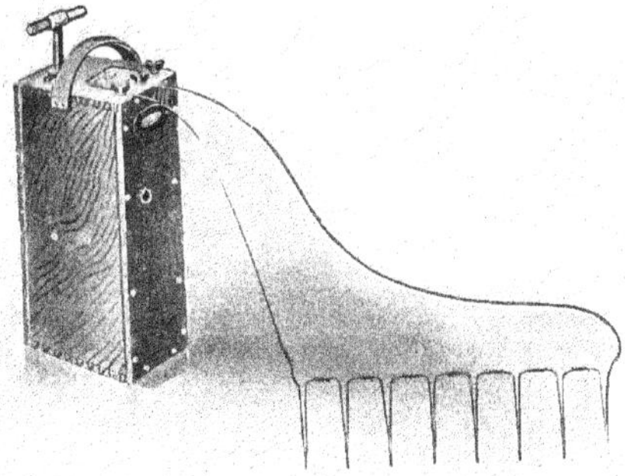

METHOD OF CONNECTING TWO LEADING WIRES TO A THREE-POST BLASTING MACHINE

METHOD OF CONNECTING THREE WIRES TO A THREE-POST BLASTING MACHINE

Leading Wire

At least two hundred and fifty feet double of leading wire should be provided, so that the blaster will be out of danger when he fires the shot. If the blast is a large one, it will probably be necessary for the blaster to be 300 or even 500 feet away. Leading wire is sold in 250 feet, 300 feet, and 500 feet coils.

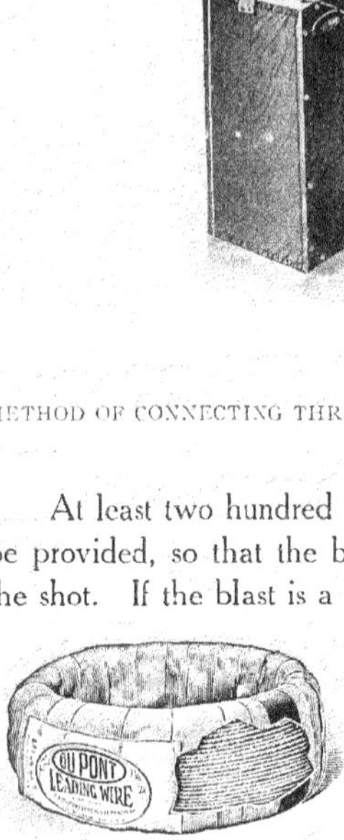

COIL OF LEADING WIRE

It is sold by the pound, and weighs about two pounds to the hundred feet.

Connecting Wire

SPOOL OF CONNECTING WIRE

Connecting wire is used to connect electric fuze wires in adjoining bore holes when the wires are not long enough to reach each other. It is wound on spools holding one pound or two pounds, and is sold by the pound.

Electric Fuzes

The electric fuze is a blasting cap having two insulated copper wires fastened into it with a composition plug. These wires are joined together in the cap by a very fine and delicate wire, which rough handling may break and make the electric fuze worthless. The charge

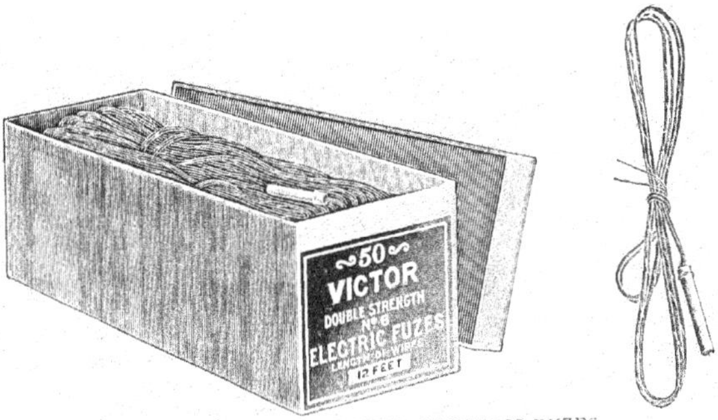

VICTOR NO. 6 (RED LABEL) ELECTRIC FUZES

which they contain is just as sensitive to shock, and just as easily affected by moisture, as that of the blasting cap; so the same rules for storing and handling must be observed. Electric fuzes are put up twenty-five or fifty in pasteboard cartons, which are packed for shipment in wooden cases. They are made in four grades: No. 4 (yellow label), No. 6 (red label), No. 7 (brown label), and No. 8 (green label), and with wires from four feet to thirty feet long. The strength of electric fuzes is exactly the same as that of blasting caps bearing the same number, and nothing weaker than the No. 6 grade can be depended on for best results from high explosives. The copper cap of the Victor No. 6 Electric Fuze is $1\frac{1}{2}$ inches to $1\frac{9}{16}$ inches long. The cartons of Victor No. 6 Electric Fuzes always have red labels on both ends.

To Prime a Cartridge with Electric Fuze

The correct way to prime a high explosive cartridge with an electric fuze, is to follow the same methods as when fuse and blasting caps are used. (See pages 100 and 102.)

The common custom of taking one or more loops, or half hitches, around the cartridge with the wires themselves, after inserting the fuze cap in a hole made diagonally in the side of the cartridge near one end, is always to be condemned. The principal objection is that the looping of the wires is very likely to break the insulation, causing short circuits, or leakage of electric current in wet work. Sometimes even the wires themselves are broken.

Ordering

Never think that because the order may be small, it is not worth while to get the very best quality of High Explosives, nor to learn how best to use them. The Du Pont Company supplies High Explosives (Dynamite), Blasting Powder, Blasting Caps, Fuse, Electric Fuzes, Blasting Machines, Wire, etc. These goods are always kept in stock. The Du Pont Company is always glad to answer any questions, and to send, free of charge, any of its illustrated catalogues on application.

If there is an uncertainty as to what grade of explosive will be required for any work, a letter, addressed to the nearest office of the Du Pont Company, telling just what work is to be done, is requested and will have the most careful attention. (See list of Branch Offices on the last page.)

If the work to be done warrants it, the Du Pont Company will be pleased to send to any part of the world at its own expense a competent man to explain the use of its explosives.

All orders for explosives, whether large or small, are filled by the Du Pont Company from its nearest stock, thus ensuring prompt and regular deliveries to its customers.

If you use explosives you want the best. Insist that your dealer sell you DU PONT GOODS AND NOTHING ELSE. They are the result of practical experience in the manufacture and use of explosives for more than a century. If your dealer does not handle DU PONT brands, write us, and we will see that your wants are supplied.

Precautions to Be Observed in General with Regard to Explosives

DON'T forget the nature of explosives, but remember that with proper care they can be handled with comparative safety.

DON'T smoke while you are handling explosives, and DON'T handle explosives near an open light.

DON'T shoot into explosives with a rifle or pistol either in or out of a magazine.

DON'T leave explosives in a field or any place where stock can get at them. Cattle like the taste of the soda and saltpetre in explosives, but the other ingredients would probably make them sick or kill them.

DON'T handle or store explosives in or near a residence.

DON'T leave explosives in a wet or damp place. They should be kept in a suitable, dry place, under lock and key, and where children or irresponsible persons cannot get at them.

DON'T explode a charge to chamber a bore hole and then immediately reload it, as the bore hole will be hot and the second charge may explode prematurely.

DON'T do tamping with iron or steel bars or tools. Use only a wooden tamping stick with no metal parts.

DON'T **force** a primer into a bore hole.

DON'T explode a charge before everyone is well beyond the danger zone and protected from flying debris. Protect your supply of explosives also from danger from this source.

DON'T hurry in seeking an explanation for the failure of a charge to explode.

DON'T drill, bore or pick out a charge which has failed to explode. Drill and charge another bore hole at least two feet from the missed one.

DON'T use two kinds of explosives in the same bore hole except where one is used as a primer to detonate the other, as where dynamite is used to detonate Judson Powder. The quicker explosive may open cracks in the rock and allow the slower to blow out through these cracks doing little or no work.

DON'T use frozen or chilled explosives. Most dynamite, except Red Cross, freezes at a temperature between 45° F. and 50° F.

DON'T use any arrangement for thawing dynamite other than one of those recommended by the DU PONT COMPANY.

DON'T thaw dynamite on heated stoves, rocks, sand, bricks or metal, or in an oven, and don't thaw dynamite in front of, near or over a steam boiler or fire of any kind.

DON'T take dynamite into or near a blacksmith shop or near a forge on open work.

DON'T put dynamite on shelves or anything else directly over steam or hot water pipes or other heated metal surface.

DON'T cut or break a dynamite cartridge while it is frozen, and don't rub a cartridge of dynamite in the hands to complete thawing.

DON'T heat a thawing house with pipes containing steam under pressure.

DON'T allow priming (the placing of a blasting cap or electric fuze in dynamite) to be done in a thawing house or magazine.

DON'T place a hot water thawer over a fire, and never put dynamite into hot water or allow it to come in contact with steam.

DON'T allow thawed dynamite to remain exposed to low temperature before using it. If it freezes again before it is used it must be thawed again.

DON'T carry blasting caps or electric fuzes in your pocket.

DON'T tap or otherwise investigate a blasting cap or electric fuze.

DON'T attempt to take blasting caps from the box by inserting a wire, nail or other sharp instrument.

DON'T try to withdraw the wires from an electric fuze.

DON'T fasten a blasting cap to the fuse with the teeth or by flattening it with a knife—use a cap crimper.

DON'T keep electric fuzes, blasting machines or blasting caps in a damp place.

DON'T attempt to use electric fuzes with the regular insulation in very wet work. For this purpose secure "Victor Water Proof" Electric Fuzes.

DON'T worry along with old, broken leading wire or connecting wire. A new supply won't cost much and will pay for itself many times over.

DON'T handle fuse carelessly in cold weather, for when cold it is stiff and breaks easily.

DON'T store or transport blasting caps or electric fuzes with high explosives.

DON'T store fuse in a hot place, as this may dry it out so that uncoiling will break it.

DON'T "lace" fuse through dynamite cartridges. This practice is frequently responsible for the burning of the charge.

DON'T operate blasting machines half-heartedly. They are built to be operated with full force. They must be kept clean and dry.

DON'T cut the fuse short to save time. It is dangerous economy.

DON'T expect a cheap article to give as good results as a high-grade one.

DON'T expect explosives to do good work if you try to explode them with a detonator weaker than a No. 6 (red label).

DU PONT

One-Pound Metal Canisters For Black Sporting Powders

DU PONT

Smokeless Shotgun Powders

METAL CANISTERS

These canisters contain "one pound bulk" of Smokeless Shotgun Powder, equivalent to 16 oz. of Black Sporting Powder

Du Pont Sporting Powders

Black and Smokeless

Best for Field and Target Shooting

When buying your powder in bulk or loaded in shells, make sure that your dealer supplies you with one of the DU PONT powders.

A day in the field after game will give you the most enjoyment if your shells are loaded with a fully guaranteed powder.

Write for information.

NOTE.—Du Pont Black Rifle Powder will not only give perfect satisfaction in your shotgun, but, if no blasting powder is to be had, Rifle Powder can be used very successfully for splitting logs, as described on page 41.

DU PONT HOME OFFICE—WILMINGTON, DEL., U. S. A.—NOVEMBER, 1910

ARE
YOU INTERESTED
IN

HIGH EXPLOSIVES

BLASTING POWDER

BLASTING SUPPLIES

SALTPETRE

BLACK SPORTING POWDER

FIREWORKS POWDER

SMOKELESS POWDER FOR SHOTGUNS

SMOKELESS POWDER FOR RIFLES

SMOKELESS POWDER FOR REVOLVERS

WOOD PULP-*WOOD FLOUR*-FOR THE MANUFACTURE OF LINOLEUM

FOR CATALOGS & READING [illegible]

WRITE NEAREST BRANCH OFFICE
(SEE LIST ON NEXT PAGE)

DU PONT

BRANCH OFFICES

BIRMINGHAM, ALA.
BOSTON, MASS.
BUFFALO, N. Y.
CHICAGO, ILL.
CINCINNATI, OHIO

DENVER, COL.
DULUTH, MINN.
HAZLETON, PA.
HOUGHTON, MICH.
HUNTINGTON, W. VA.

JOPLIN, MO.
KANSAS CITY, MO.
MEMPHIS, TENN.
MEXICO CITY, MEXICO
NASHVILLE, TENN.

NEW ORLEANS, LA.
NEW YORK, N. Y.
PHILADELPHIA, PA.
PITTSBURG, KAS.
PITTSBURGH, PA.

PORTLAND, ORE.
SALT LAKE CITY, UTAH
SAN FRANCISCO, CAL.
SCRANTON, PA.
SEATTLE, WASH.

SHREVEPORT, LA.
SPOKANE, WASH.
SPRINGFIELD, ILL.
ST. LOUIS, MO.
TERRE HAUTE, IND.

Lc Ap '11

Spotted something that isn't quite right?

Let us know and we'll fix it for future reprints.
And as a thank you for your diligence, we'll
send you a free Sothis book of your choice.

publisher@sothis.press

This page intentionally left blank

Except of course for that line above. And this one.
And whilst we have your attention, perhaps you'd
like to follow us on Instagram @sothis.press. Go on.

Made in the USA
Monee, IL
05 November 2026

67808943R00080